FEARFULLY
AND
WONDERFULLY
MADE

Books available from Marshall Pickering

By the same authors

IN HIS IMAGE

By Philip Yancey

WHERE IS GOD WHEN IT HURTS?

FEARFULLY
AND
WONDERFULLY
MADE

Philip Yancey
and
Dr Paul Brand

Illustrated by Charles Shaw

MarshallPickering
An Imprint of HarperCollins*Publishers*

Marshall Pickering is an imprint of
HarperCollins*Religious*
Part of HarperCollins*Publishers*
77 85 Fulham Palace Road,
Hammersmith, London W6 8JB

First published in Great Britain by
Marshall Pickering in 1993
1 3 5 7 9 8 6 4 2

First published in the USA in 1980 by
Zondervan Books
Paperback edition 1987

A catalogue record for this book is
available from the British Library

ISBN 0 551 02322 8

Set in Baskerville

Printed in Great Britain

Men go abroad to wonder at the height of mountains, at the huge waves of the sea, at the long courses of the rivers, at the vast compass of the ocean, at the circular motion of the stars; and they pass by themselves without wondering.

SAINT AUGUSTINE

You created my inmost being; you knit me together in my mother's womb. I praise you because I am fearfully and wonderfully made.

DAVID
Psalm 139:13–14

Contents

In the process of writing this book, some twenty people gave us valuable editorial comments and suggestions, for which we are profoundly grateful. Three in particular – Harold Fickett, Elizabeth Sherrill, and Tim Stafford – offered constructive comments which led to a major restructuring of the entire manuscript. We offer special thanks to them and to our perceptive and faithful editor at Zondervan, Judith Markham.

Preface

Sometimes when uncertain of a voice from its very loudness, we catch the missing syllable in the echo. In God and Nature we have Voice and Echo.

HENRY DRUMMOND

Except in this preface, the personal pronoun "I" will always refer to Dr Paul Brand; the book is written from his perspective. Yet, unlike many books with co-authors, this was not written in an "as told to" style.

I first met Dr Brand while doing research for my book *Where Is God When It Hurts?* His medical credentials in the field of pain are unquestioned. Besides eighteen years of brilliant pioneering research on the disease of leprosy in India, he has attained world stature as a hand surgeon and rehabilitation specialist. In honour of his contributions he has received the prestigious Albert Lasker Award and has been made Commander of the Order of the British Empire by Queen Elizabeth.*

I knew these facts about Dr Brand before visiting him at the leprosy hospital in Carville, Louisiana, where he works and lives, but I did not know the degree to which his Christian faith has permeated his life and thought. As an avid scientist, bird watcher, mountain climber, and organic gardener, he has striven to integrate the natural order with the spiritual order.

* The biography *Ten Fingers for God* narrates his life story.

During my second visit, Dr Brand hesitantly pulled out a ninety-page manuscript, part typewritten, part doctor's scratchings, which contained some of his thoughts on the human body. He had developed the manuscript from talks given at the Christian Medical College in Vellore, India. "In a sense," he said, "we doctors are like employees at the complaint desk of a large department store. We tend to get a biased view of the quality of the product when we hear about its aches and pains all day. In this little manuscript, which I set aside long ago, I tried to pause and wonder at what God made. I took an old analogy from the New Testament and updated it with the expanded knowledge we've gained from modern science. Curiously, every medical discovery seems to make the analogy fit even better; not one has weakened the original meanings the apostle Paul set forth."

The idea of a book on the body analogy attracted me because I, too, appreciate the harmony between the natural and spiritual worlds.

G. K. Chesterton's book *St Francis of Assisi* proposes the intriguing theory that the Dark Ages resulted because paganism and mythologies had stained the natural order so deeply that Christians could not perceive nature as part of God's revelation. "It was no good telling such people to have a natural religion full of stars and flowers; there was not a flower or even a star that had not been stained. They had to go into the desert where they could find no flowers or even into the cavern where they could see no stars."[1] As a result, virtually all forms of art sank to a low level during this period of civilization. For the Christians, nature had been cleaved from supernature.

Today, a similar process is taking place. The created world has lost its sacredness. Christians have abandoned it, not to paganism, but to physics, geology, biology and chemistry. We too have cleaved nature from the supernatural.

I saw in Dr Brand a man with impeccable credentials in his field of science, yet one who also could contribute a humble awareness of how nature echoes its Creator. We worked out many of the specific applications together, and then I spent several months researching the medical backgrounds of each of the analogies. After long hours of interviewing, I also managed to penetrate Dr Brand's modesty and British reserve to tap into his vast reservoir of dramatic personal experiences.

Dr Brand and I desire that this book will help span the chasm that for too long has separated the created world from its Source. God invented matter. He invested His great, creative self in this world and, specifically, in the design of our bodies. The least we can do is be grateful.

Hopefully, the book will also offer insight into the mysterious, organic relationship that exists among the people of God. New Testament writers kept drifting back to a single metaphor to express this relationship: the Body of Christ. In this first book together we examine the Body and four of its parts; more may follow in a future volume.

Someone attempting to describe the colour scarlet to a person born blind declared, "It's like the sound of a trumpet". In a sense, metaphorical symbols are the only way for us to grasp spiritual truths, which explains why the Bible uses them so lavishly. And symbols have a certain power. As John V. Taylor says, " 'No man is an island' has fifty times more voltage than 'No man is self-sufficient.' " In the field of religious publishing dominated by books heavy on theological content or personal experience, we hope these analogies provide another mode of perceiving reality.

If it ever bothers you that we explore the body analogy

more fully than the Bible does, please close the book. We do not want to bend truth to fit an analogy. On the other hand, you may find, as we did, that the human body expresses spiritual reality so authentically that soon the common stuff of matter will appear more and more like a mere shadow.

PHILIP YANCEY

CELLS

1
MEMBERS

*I have been trying to think of the earth as a kind of organism,
but it is no go. I cannot think of it this way. . . . The other
night, driving through a hilly, wooded part of southern New
England, I wondered about this. If not like an organism,
what is it like, what is it most like? Then, satisfactorily for
that moment, it came to me: it is most like a single cell.*

LEWIS THOMAS

I remember the first time I saw a living cell under a
microscope. I was twenty-one years old and taking a short
course in tropical hygiene at Livingstone College in
England. We had been studying parasites, but our
specimens were dead; I wanted to see a living amoeba.
Early one morning, before the laboratory was cluttered with
students, I sneaked into the old science building. The
imposing red brick structure stood next to a pond from
which I had just scooped some water in a teacup. Bits of
decomposing leaves floated in the turbid water, smelling
of decay and death.

But when I touched one drop of that water to a
microscope slide, a universe sprang to life. Hundreds of
organisms crowded into view: delicate, single-celled globes
of crystal, breathing, unfurling, flitting sideways, excited
by the warmth of my microscope light. I edged the slide
a bit, glancing past the faster organisms. Ah, there it was.

An amoeba. A mere chip of translucent blue, it was barely visible to my naked eye, but the microscope revealed even its inner workings.

Something about the amoeba murmurs that it is one of the most basic and primordial of all creatures. Somehow it has enlisted the everyday forces of millions of spinning atoms so that they now serve life, which differs profoundly from mere matter. Just an oozing bit of gel, the amoeba performs all the basic functions that my body does. It breathes, digests, excretes, reproduces. In its own peculiar way it even moves, plumping a hummock of itself forward and following with a motion as effortless as a drop of oil spreading on a table. After one or two hours of such activity, the grainy, watery blob will have travelled a third of an inch.

That busy, throbbing drop gave me my first graphic image of the jungle of life and death we share. I saw the amoeba as an autonomous unit with a fierce urge to live and a stronger urge to propagate itself. It beckoned me on to explore the living cell.

*

Years later I am still observing cells, but as a physician I focus on how they cooperate within the body.

Now I have my own laboratory, at a leprosy hospital on swampy ground by the Mississippi River in Carville, Louisiana. Again I enter the lab. early before anyone is stirring, this time on a chilly winter morning. Only the soft buzz of fluorescent lights overhead breaks the quietness.

But I have not come to study amoebae. This morning I will examine a hibernating albino bat who sleeps in a box in my refrigerator. I rely on him to study how the body responds to injury and infection. I lift him carefully, lay him on his back, and spread his wings in a cruciform posture. His face is weirdly human, like the shrunken heads in museums. I keep expecting him to open an eye and shriek at me, but he doesn't. He sleeps.

As I place his wing under the microscope lens, again a new universe unfolds. I have found a keyhole. The albinic skin under his wing is so pale that I can see directly through his skin cells into the pulsing capillaries which carry his blood. I focus the microscope on one bluish capillary until I can see individual blood cells pushing, blocking, thrusting through it. Red blood cells are by far the most numerous: smooth, shiny discs with centres indented like jelly doughnuts. Uniform size and shape make them seem machine-stamped and impersonal.

More interesting are the white blood cells, the armed forces of the body which guard against invaders. They look exactly like the amoebae: amorphous blobs of turgid liquid with darkened nuclei, they roam through the bat's body by extending a finger-like projection and humping along to follow it. Sometimes they creep along the walls of the veins; sometimes they let go and free-float in the bloodstream. To navigate the smaller capillaries, bulky white cells must elongate their shapes, while impatient red blood cells jostle in line behind them.

Watching the white cells, one can't help thinking them sluggish and ineffective at patrolling territory, much less repelling an attack. Until the attack occurs, that is. I take a steel needle and, without waking the bat, prick through its wing, puncturing a fine capillary. An alarm seems to sound. Muscle cells contract around the damaged capillary wall, damming up the loss of precious blood. Clotting agents halt the flow at the skin's surface. Before long, scavenger cells appear to clean up debris, and fibroblasts, the body's reweaving cells, gather around the injury site. But the most dramatic change involves the listless white cells. As if they have a sense of smell (we still don't know how they "sense" danger), nearby white cells abruptly halt their aimless wandering. Like beagles on the scent of a rabbit, they home in from all directions to the point of attack. Using their unique shape-changing qualities, they ooze between overlapping cells of capillary walls and hurry through tissue via the most direct route. When they arrive, the battle begins.

Lennart Nilsson, the Swedish photographer famous for his remarkable closeups of activity inside the body, has captured the battle on film as seen through an electron microscope. In the distance, a shapeless white cell, resembling science fiction's creature "The Blob", lumbers toward a cluster of luminous green bacterial spheres. Like a blanket pulled over a corpse, the cell assumes their shape; for a while they still glow eerily inside the white cell. But the white cell contains granules of chemical explosives, and as soon as the bacteria are absorbed the granules detonate, destroying the invaders. In thirty seconds to a minute only the bloated white cell remains. Often its task is a kamikaze one, resulting in the white cell's own death.

In the body's economy, the death of a single white cell is of little consequence. Most only live several days or several weeks, and besides the fifty billion active ones prowling the adult human, a backup force one hundred times as large lies in reserve in the bone marrow. At the cellular level, massive warfare is a daily fact of life. Fifty thousand invaders may lurk on the rim of a drinking glass, and a billion can be found in a half-teaspoon of saliva. Bacteria enshroud my body – every time I wash my hands I sluice five million of them from the folds of my skin.*

To combat these threats, some of the blood's white cells are specifically targeted to one type of invader. If the body has experienced contact with a severe danger, as in a smallpox vaccination, it imprints certain white cells with a death-wish to combat that signal danger. These cells spend their lives coursing through the bloodstream, waiting, scouting. Often they are never called upon to give battle. But if they are, they hold within them the power to disarm a foreign agent that could cause the destruction of every cell in the body.

*

Often I have reflected on the paradox of the amoeba and its mirror image, the white cell. The amoeba, a self-contained organism, alone performs all the basic functions of life, depending on other cells only when it ingests them as food. The white cell, though similar in construction and makeup, in a sense is far less free. A larger organism determines its duties, and it must sometimes sacrifice its life for the sake of that organism. Although more limited in self-expression, the white cell performs a singularly vital function. The amoeba flees danger; the white cell moves toward it. A white cell can keep alive a person like Beethoven or Newton or Einstein . . . or you and me.

I sometimes think of the human body as a community, and then of its individual cells such as the white cell. The cell is the basic unit of an organism; it can live for itself, or it can help form and sustain the large organism. I recall the apostle Paul's use of analogy in 1 Corinthians 12 where he compares the church of Christ to the human body. That inspired analogy takes on even more meaning to me because

---

* The quantities of bacteria viewed through the first effective microscopes so overwhelmed scientists that subsequent generations have lived in vivid awareness of "germs". Astute promoters market disinfectants to sterilize our environment, but too often the germ-killer, merely a cell-killer, also destroys the body's good cells. Today we need better publicity for our bodies' able defences and perhaps less fear of germs – the average American household is in more danger from chemical germ-killers than from germs. I prefer to leave the battle to my own cells.

of the expanded vistas allowed by the invention of microscopes. Since Paul's analogy renders a basic principle of God's creation, I can augment it like this:

> The body is one unit, though it is made up of many cells, and though all its cells are many, they form one body. . . . If the white cell should say, because I am not a brain cell, I do not belong to the body, it would not for that reason cease to be part of the body. And if the muscle cell should say to the optic nerve cell, because I am not an optic nerve cell, I do not belong to the body, it would not for that reason cease to be part of the body. If the whole body were an optic nerve cell, where would be the ability to walk? If the whole body were an auditory nerve cell, where would be the sense of sight? But in fact God has arranged the cells in the body, every one of them, just as He wanted them to be. If all cells were the same, where would the body be? As it is, there are many cells, but one body.

That analogy conveys a more precise meaning to me because though a hand or foot or ear cannot have a life separate from the body, a cell does have that potential. It can be part of the body as a loyalist, or it can cling to its own life. Some cells do choose to live in the body, sharing its benefits while maintaining complete independence – they become parasites or cancer cells.

2

SPECIALIZATION

To be a member is to have neither life, being, nor movement,
except through the spirit of the body, and for the body.
BLAISE PASCAL

The scientist who collects and catalogues and the child who wanders barefoot through the woods are equally awestruck by the sheer profusion of creatures that populate this planet. The child marvels at the psychedelic design of a butterfly; he chases darting "skeeter hawks" (dragonflies), yelps at the spastic leap of a click beetle, breathlessly fondles a baby rabbit. The scientist looks closer. He takes a simple block of forest soil, one foot square and one inch deep, and begins counting. In this loamy world that we so thoughtlessly tread upon he finds "an average of 1,356 living creatures, including 865 mites, 265 springtails, 22 millipedes, 19 adult beetles, and various numbers of 12 other forms."[1] Without an electron microscope and infinite patience he cannot bother with the two billion bacteria and the millions of fungi and algae.

In his laboratory, the scientist begins with our friend the amoeba and works up, classifying from the "lower" to the "higher". What is this term "lower"? How can we trample a million creatures on a hike and return home guiltless? A devout vegetarian who gulps cold spring water imbibes a horde of creatures – animals! – without flinching. Why do we wince at a bloodied cat along the roadside but take no notice of the billions of tiny animals pulverized by the bulldozer scraping out a roadbed?

The key to our ranking of worth is specialization: the process of cells taking turns, dividing up labour, limiting their responses to a single task. We recognize a more meaningful life in the cat, a higher animal consisting of lower cells working together. The amoeba on my microscope slide is the animal at the bottom of the zoological ladder. It moves, yes, but barely an inch a day. It may spend its lifetime confined to a tin can or the hollowed part of an old tyre. Unlike some humans, it will never tour Europe, visit the Taj Mahal, climb the Rockies. To do that one needs specialized muscle cells, rows and rows of them, aligned like stalks of wheat. The lower animals skitter, creep, or worm along, covering mere yards of turf. The higher ones hop and leap and gallop. Others, winged creatures, vault and soar and dive. It is a matter of specialization.

Consider the organ of sight. An amoeba has some crude visual awareness: it moves toward light – and that is all. Specialization gives the human the ability to be on the viewing end of the microscope, eyeing the subtleties of colour in the near-senseless amoeba. The amoeba has one cell. Inside my human eye, peering at him, are 107,000,000 cells. Seven million are cones, each loaded to fire off a message to the brain when a few photons of light cross them. Cones give me the full band of colour awareness, and because of them I can easily distinguish a thousand shades of colour. The other hundred million cells are rods, backup cells for use in low light. When only rods are operating, I do not see colour (as on a moonlit night when everything looms in shades of grey, but I can distinguish a spectrum of light so broad that the brightest light I perceive is a billion times brighter than the dimmest.

Between the amoeba and my eye exists a boggling range of specialization. The copilia, an animal living in the Bay of Naples, has only one visual receiver, a cone cell attached to a muscular stalk which scans like a roving television camera. Although the copilia can absorb only one light message at a time, presumably its brain can merge many messages into a crude picture of its environment.

The human brain receives millions of simultaneous reports from eye cells. If its designated wavelength of light is present, each rod or cone triggers an electrical response to the brain. The brain then absorbs a composite set of yes or no messages from all the rods and cones. It sorts and organizes them all and gives me an image of an amoeba swimming on my microscope slide. Compared to the amoeba's one-celled independence, the stationary lives of my rods and cones appear drab indeed. But who among us would trade ends of the eyepiece?

For specialization to work, the individual cell must lose all but one or two of its abilities. A rod or cone cell cannot move about, while an amoeba can perform a whole array of minuscule activities. But the human cell can, through its limited role, make possible much "higher", more meaningful achievement. A single rod can provide me with the wavelength of light that completes my appreciation of a rainbow, a kingfisher plunging into a stream, or a subtle change of expression in the face of a dear friend. Or it may protect me from disaster by firing off a message to the brain when a rock is hurled from an expressway bridge at my approaching car.

*

In exchange for its self-sacrifice, the individual cell can share in what I call the ecstasy of community. No scientist can yet measure how a sense of security or pleasure is communicated to the cells of the body, but individual cells certainly participate in our emotional reactions. Hormones and enzymes bathe them, bringing on a quickened breathing, a trembling of muscles, a flapping in the stomach. If you look for a pleasure nerve in the human body, you will come away disappointed; there is none. There are nerves for pain and cold and heat and touch, but no nerve gives a sensation of pleasure. Pleasure appears as a by-product of cooperation by many cells.

What about sexual pleasure? Even that is not as specific and localized as you may think. Erogenous zones have no specialized pleasure nerves; the cells concentrated there also sense touch and pain. Besides the touch stimulation of skin against skin, sex includes a sense of need and visual delights, memories, and perhaps the auditory stimulus of background music. We also bring to sex that complex, compulsive love that loves oneself and another at the same moment. At a still deeper, cellular level lies an urge to propagate life, to ensure survival, which is programmed into every cell. All these factors work together to produce sexual pleasure.

I greatly enjoy another human pleasure: listening to a symphony orchestra. When I do, the chief source of what I interpret as pleasure is located inside my ear. There I can detect sound frequencies that flutter my eardrums as faintly as one billionth of a centimetre (a distance one-tenth the diameter of a hydrogen atom). This vibration is transmitted into my inner ear by three bones familiarly known as the hammer, anvil, and stirrup. When the frequency of middle C is struck on a piano, the piston of bones in my inner ear vibrates 256 times a second. Further in are individual cilia, comparable to the rods and cones of the eye, that transmit specific messages of sound to the

brain. My brain combines these messages with other factors
– how well I like classical music, how familiar I am with
the piece being played, the state of my digestion, the friends
I am with – and offers the combination of impulses in a
form I perceive as pleasure.

Nature includes some organisms who cooperate but
cannot quite pull off this ecstasy of community. For
example, certain strains of amoebae come together for the
purpose of reproduction. These ''social amoebae,'' as few
as ten and as many as five hundred thousand, join in a
short-lived phenomenon called a slime mould. They group
in an orderly fashion, forming a tiny, lustrous, bullet-shaped
slug. As the slug inches forward, it leaves a trail of slime,
hence the name. The cells in front cooperate until a tower
of them reaches into the air. At the top forms an orbic spore
full of amoebae, giving the slug a new shape, almost like
a toadstool. Suddenly the spore explodes, scattering new
amoebae into the environment. The whole phenomenon
takes eight hours and demonstrates a simple form of cooperation among the cells. Yet in the slime-mould amoebae
something is lacking. At no point is there only one organism
imprinted with the same genes and the same loyalties. Many
slime-mould cells cooperate for the single event of
reproduction, then break off and go their own way.

In contrast, the human body grows from the fertilization
of a single egg. In *The Medusa and the Snail*, Lewis Thomas
muses about why people made such a fuss over the test-
tube baby in England. The true miracle, he affirms, is the
common union of a sperm and egg in a process that
ultimately produces a human being. ''The mere existence
of that cell,'' he writes, ''should be one of the greatest
astonishments of the earth. People ought to be walking
around all day, all through their waking hours, calling to
each other in endless wonderment, talking of nothing except

that cell. . . . If anyone does succeed in explaining it, within my lifetime, I will charter a skywriting aeroplane, maybe a whole fleet of them, and send them aloft to write one great exclamation point after another, around the whole sky, until all my money runs out.''[2]

Over nine months these cells divide up functions in exquisite ways. Billions of blood cells appear, millions of rods and cones – in all, up to one hundred million million cells form from a single fertilized ovum. And finally a baby is born, glistening with liquid. Already his cells are cooperating. His muscles limber up in jerky, awkward movements; his face recoils from the harsh lights and dry air of the new environment; his lungs and vocal chords join in a first air-gulping yell.

Within that clay-coloured, wrinkled package of cells, lies the miracle of the ecstasy of community. His life will include the joy of seeing his mother's approval at his first clumsy words, the discovery of his own unique talents and gifts, the fulfilment of sharing with other humans. He is many cells, but his is one organism. All of his hundred trillion cells know that.

*

I have closed my eyes. My shoes are kicked off, and I am wiggling the small bones in my right foot. Exposed, they are half the width of a pencil, and yet they support my weight in walking. I cup my hand over my ear and hear the familiar seashell phenomenon, actually the sound of blood cells rushing through the capillaries in my head. I stretch out my left arm and try to imagine the millions of muscle cells eagerly expanding and contracting in concert. I rub my finger across my arm and feel the stimulation of touch cells, 450 of them in each one-inch-square patch of skin.

Inside, my stomach, spleen, liver, pancreas, and kidneys, each packed with millions of loyal cells, are working so efficiently I have no way of perceiving their presence. Fine hairs in my inner ear are monitoring a swishing fluid, ready to alert me if I suddenly tilt off balance.

When my cells work well, I'm hardly conscious of their individual presences. What I feel is the composite of their activity known as Paul Brand. My body, composed of many parts, is one. And that is the root of the analogy we shall explore.

3

DIVERSITY

*We often think that when we have completed our study on
one, we know all about two, because "two is one and one."
We forget that we have still to make a study of "and."*
                                      SIR ARTHUR EDDINGTON

More than amoebae and bats skulk in my medical
laboratory. One drawer contains neatly filed specimens of
an array of cells from the human body. Separated from
the body, lifeless, stained with dyes and mounted in epoxy,
they hardly express the churn of living cells at work inside
me at this moment. But if I parade them under the
microscope, certain impressions about the body take shape.

I am first struck by their variety. Chemically my cells
are almost alike, but visually and functionally they are as
different as the animals in a zoo. Red blood cells, discs
resembling Lifesaver candies, voyage through my blood
loaded with oxygen to feed the other cells. Muscle cells,
which absorb so much of that nourishment, are sleek and
supple, full of coiled energy. Cartilage cells with shiny black
nuclei look like bunches of black-eyed peas glued tightly
together for strength. Fat cells seem lazy and leaden, like
bulging white plastic garbage bags jammed together.

Bone cells live in rigid structures that exude strength.
Cut in cross section, bones resemble tree rings, overlapping

strength with strength, offering impliability and sturdiness. In contrast, skin cells form undulating patterns of softness and texture that rise and dip, giving shape and beauty to our bodies. They curve and jut at unpredictable angles so that every person's fingerprint – not to mention his or her face – is unique.

The aristocrats of the cellular world are the sex cells and nerve cells. A woman's contribution, the egg, is one of the largest cells in the human body, its ovoid shape just visible to the unaided eye. It seems fitting that all the other cells in the body should derive from this elegant and primordial structure. In great contrast to the egg's quiet repose, the male's tiny sperm cells are furiously flagellating tadpoles with distended heads and skinny tails. They scramble for position as if competitively aware that only one of billions will gain the honour of fertilization.

The king of cells, the one I have devoted much of my life to studying, is the nerve cell. It has an aura of wisdom and complexity about it. Spider-like, it branches out and unites the body with a computer network of dazzling sophistication. Its axons, "wires" carrying distant messages to and from the human brain, can reach a yard in length.

I never tire of viewing these varied specimens or thumbing through books which render cells. Individually they seem puny and oddly designed, but I know these invisible parts cooperate to lavish me with the phenomenon of life. Every second my smooth muscle cells modulate the width of my blood vessels, gently push matter through my intestines, open and close the plumbing in my kidneys. When things are going well – my heart contracting rhythmically, my brain humming with knowledge, my lymph laving tired cells – I rarely give these cells a passing thought.

But I believe these cells in my body can also teach me

about larger organisms: families, groups, communities, villages, nations — and especially about one specific community of people that is likened to a body more than thirty times in the New Testament. I speak of the Body of Christ, that network of people scattered across the planet who have little in common other than their membership in the group that follows Jesus Christ.

My body employs a bewildering zoo of cells, none of which individually resembles the larger body. Just so, Christ's Body comprises an unlikely assortment of humans. Unlikely is precisely the right word, for we are decidedly unlike one another and the One we follow. From whose design come these comical human shapes which so faintly reflect the ideals of the Body as a whole?

*

Novelist Frederick Buechner playfully described the motley crew God selected in Bible times to accomplish His work:

"Who could have predicted that God would choose not Esau, the honest and reliable, but Jacob, the trickster and heel, that he would put the finger on Noah, who hit the bottle, or on Moses, who was trying to beat the rap in Midian for braining a man in Egypt and said if it weren't for the honour of the thing he'd just as soon let Aaron go back and face the music, or on the prophets, who were a ragged lot, mad as hatters most of them. . . . ?

"And of course, there is the comedy, the unforeseeableness, of the election itself. Of all the peoples he could have chosen to be his holy people, he chose the Jews, who as somebody has said are just like everybody else only more so — more religious than anybody when they were religious and when they were secular, being secular as if they'd invented it. And the comedy of the covenant — God saying 'I will be your God and you shall be my people' (Exodus 6:7) to a people who before the words had stopped ringing in their ears were dancing around the golden calf like aborigines and carrying on with every agricultural deity and fertility god that came down the track."[1]

The exception seems to be the rule. The first humans God created went out and did the only thing God asked them not to do. The man he chose to head a new nation known as "God's people" tried to pawn off his wife on an unsuspecting Pharaoh. And the wife herself, when told at the ripe old age of ninety-one that God was ready to deliver the son He has promised her, broke into rasping laughter in the face of God. Rahab, a harlot, became revered for her great faith. And Solomon, the wisest man who ever lived, went out of his way to break every proverb he so astutely composed.

Even after Jesus came the pattern continued. The two disciples who did most to spread the word after His departure, John and Peter, were the two He had rebuked most often for petty squabbling and muddleheadedness. And the apostle Paul, who wrote more books than any other Bible writer, was selected for the task while kicking up dust whirls from town to town sniffing out Christians to torture. Jesus had nerve, entrusting the high-minded ideals of love and unity and fellowship to this group. No wonder cynics have looked at the church and sighed, "If that group of people is supposed to represent God, I'll quickly vote against Him." Or, as Nietzsche expressed it, "His disciples will have to look more saved if I am to believe in their saviour."

Yet our study of the Body of Christ must allow for this impossible dream, for all we are is a collection of people as diverse as the cells in the human body. I think of the churches I have known. Is there another institution in town with such a mosaic assortment of unlikes? Young radicals, uniformed in jeans, share the pews with Republican bankers in three-piece suits. Bored teenagers switch off from the sermon even as their eager grandparents turn up their hearing aids. Some members gather as methodically as a school of fish, then quickly break apart to return to their jobs and homes. Others want close communities and migrate together like social amoebae.

I could easily cluck my tongue at the absurdity of the whole enterprise, seemingly doomed to fail. Jesus prayed that we "may be one" as He and God the Father are one. (John 17:11) How can any organism composed of such diversity attain even a semblance of unity?

As the doubts rumble inside me, a sober and quieting voice replies, "You have not chosen Me. I have chosen you." The chuckle at Christ's Body is caught in my throat like cotton. For if anything is to be believed about the collection of people who follow Him, it is that we were called by Him. The word church, *ekklēssia*, means "the called-out ones". Our crews of comedians from central casting is the group God wants.

During my life as a missionary surgeon in India and now as a member of the tiny chapel on the grounds of the Carville leprosy hospital, I have seen my share of unlikely seekers after God. And I must admit that most of my worship in the last thirty years has not taken place among people who have shared my tastes in music, speech, or even thought. But over those years I have been profoundly – and humbly – impressed that I find God in the faces of my fellow worshippers by sharing with people who are shockingly different from each other and from me.

C. S. Lewis recounts that when he first started going to church he disliked the hymns, which he considered to be fifth-rate poems set to sixth-rate music. But as he continued, he said, "I realized that the hymns (which were just sixth-rate music) were, nevertheless, being sung with devotion and benefit by an old saint in elastic-side boots in the opposite pew, and then you realize that you aren't fit to clean those boots. It gets you out of your solitary conceit."[2]

A colour on a canvas can be beautiful in itself. However, the artist excels not by slathering one colour across the canvas but by positioning it between contrasting or complementary hues. The original colour then derives richness and depth from its milieu of unlike colours.

The basis for our unity within Christ's Body begins not with our similarity but with our diversity.

*

It seems safe to assume that God enjoys variety, and not just at the cellular level. He didn't stop with a thousand insect species; He conjured up three hundred thousand species of beetles and weevils alone. In his famous speech in the Book of Job, God pointed with pride to such oddities of creation as the mountain goat, the wild ass, the ostrich, and the lightning bolt. He lavished colour, design and texture on the world, giving us Pygmies and Watusis, blond Scandinavians and swarthy Italians, big-boned Russians and petite Japanese.

People, created in His image, have continued the process of individualization, grouping themselves according to distinct cultures. Consider the continent of Asia for a crazy salad. In China women wear long pants and men wear gowns. In tropical Asia people drink hot tea and munch on blistering peppers to keep cool. Japanese fry ice cream. Indonesian men dance in public with other men to demonstrate that they are not homosexual. Westerners smile at the common Asian custom of marriages arranged by parents; Asians gasp at our entrusting such a decision to vague romantic love. Balinese men squat to urinate and women stand. Many Asians begin a meal with a sweet and finish it with a soup. And when the British introduced the violin to India a century ago, men started playing it while sitting on the floor, holding it between the shoulder and the sole of the foot. Why not?

Whenever I travel overseas, I am struck anew by the world's incredible diversity, and the churches overseas are now beginning to show that cultural self-expression. For too long they were bound up in Western ways (as the early church had been bound in Jewish ways) so that hymns, dress, architecture, and church names were the same around the world. Now indigenous churches are bursting out with their own spontaneous expressions of worship to God. I must guard against picturing the Body of Christ as composed only of American or British cells; it is far grander and more luxuriant than that.

I grew up in a denomination called the Strict and Particular Baptists, from which I learned faith and love for God and the Bible. Unfortunately, I also was taught how crucially better we were than every other church. We were not even allowed to have communion with other Baptist denominations. My great-grandparents, Huguenots, had escaped Catholic persecution in France, and as children we were taught that nuns and priests were akin to the devil. My Christian growth since those days has required some abrupt adjustments.

I have learned that when God looks upon His Body, spread like an archipelago throughout the world, He sees the whole thing. And I think He, understanding the cultural backgrounds and true intent of the worshippers, likes the variety He sees.

Blacks in Murphy, North Carolina, shout their praises to God. Believers in Austria intone them, accompanied by magnificent organs and illuminated by stained glass. Some Africans dance their praise to God, following the beat of a skilled drummer. Sedate Japanese Christians express their gratitude by creating objects of beauty. Indians point their hands upward, palms together, in the *namaste* greeting of respect, that has its origin in the Hindu concept, "I worship the God I see in you", but gains new meaning as Christians use it to recognize the image of God in others.

The Body of Christ, like our own bodies, is composed of individual, unlike cells that are knit together to form one Body. He is the whole thing, and the joy of the Body increases as individual cells realize they can be diverse without becoming isolated outposts.

4
WORTH

Whereas American mothers preserve, often in fronze, their children's first shoes – celebrating freedom and independence – a Japanese mother carefully preserves a small part of her child's umbilical cord – celebrating dependence and loyalty.

STEPHEN FRANKLIN

As a boy growing up in India, I idolized my missionary father who responded to every human need he encountered. Only once did I see him hesitate to help – when I was seven, and three strange men trudged up the dirt path to our mountain home.

At first glance these three seemed like hundreds of other strangers who streamed to our home for medical treatment. Each was dressed in a breechcloth and turban, with a blanket draped over one shoulder. But as they approached, I noticed differences: a mottled quality to their skin, thick, swollen foreheads and ears, and strips of blood-stained cloth bandaging their feet. As they came closer, I noticed they also lacked fingers and one had no toes – his limbs ended in rounded stumps.

My mother's reaction differed from her normal gracious hospitality. Her face took on a pale, tense appearance. "Run and get papa," she whispered to me. "Take your sister, and both of you stay in the house!"

My sister obeyed perfectly, but after calling my father
I scrambled on hands and knees to a nearby vantage point.
Something sinister was happening, and I didn't want to
miss it. My heart pounded violently as I saw the same look
of uncertainty, almost fear, pass across my father's face.
He stood by the three nervously, awkwardly, as if he didn't
know what to do. I had never seen my father like that.

The three men prostrated themselves on the ground, a
common Indian action that my father disliked. "I am not
God – He is the One you should worship," he would
usually say, and lift the Indians to their feet. But not this
time. He stood still. Finally, in a weak voice he said,
"There's not much we can do. I'm sorry. But wait where
you are; don't move. I'll do what I can."

He ran to the dispensary while the men squatted on the
ground. Soon he returned with a roll of bandages, a can
of salve, and a pair of surgical gloves he was struggling to
put on. This was most unusual – how could he treat them
wearing gloves?

Father washed the strangers' feet, applied ointment to
their sores, and bandaged them. Strangely, they did not
wince or cry out as he touched their sores.

While father bandaged the men, mother had been
arranging a selection of fruit in a wicker basket. She set
it on the ground beside them, suggesting they take the
basket. They took the fruit but left the basket, and as they
disappeared over the ridge I went to pick it up.

"No!" mother insisted. "Don't touch it! And don't go
near that place where they sat." Silently I watched father
take the basket and burn it, then scrub his hands with hot
water and soap. Then mother bathed my sister and me,
though we had had no direct contact with the visitors.

That incident was my first exposure to leprosy, the oldest
recorded disease and probably the most dreaded disease

throughout history. Although I might have recoiled from the suggestion as a boy of seven, I eventually felt called to spend my life working among leprosy patients. For the past thirty years I have been with them almost daily, forming many intimate and lasting friendships among these courageous people. During that time, many exaggerated fears and prejudices about leprosy have crumbled, at least in the medical profession. Partly because of effective drugs, leprosy is now viewed as a controllable, barely contagious disease.

However, in most parts of the world less than a quarter of leprosy patients are actually under any form of treatment. Thus, to many it is still a disease that can cause severe lesions, blindness, and loss of hands and feet. How does leprosy produce such terrible effects?

*

As I studied leprosy patients in India, several findings pushed me toward a rather simple theory: could it be that the horrible results of the disease came about because leprosy patients had lost the sense of pain? The disease was not at all like a flesh-devouring fungus; rather, it attacked mainly a single type of cell, the nerve cell. After years of testing and observation, I felt sure that the theory was sound.

The gradual loss of the sense of pain leads to misuse of those body parts most dependent on pain's protection. A person uses a hammer with a splintery handle, does not feel the pain, and an infection flares up. Another steps off a kerb, spraining an ankle, and, oblivious, keeps walking. Another loses use of the nerve that triggers the eyelid to blink every few seconds for lubricating moisture; the eye dries out, and the person becomes blind.

The millions of cells in a hand or foot, or the living and

alert rod and cone cells in the eye, can be rendered useless because of the breakdown of just a few nerve cells. Such is the tragedy of leprosy.

A similar pattern can be found in other diseases. In sickle cell anaemia or leukaemia the malfunction of a single type of cell can quickly destroy a person. Or, if the cells that keep kidney filters in repair fail, a person may soon die of toxic poisoning.

This fact of the body – the worth of each of its parts – is graphically revealed by a disease such as leprosy. The failure of one type of cell can bring on tragic consequences. One who studies the vast quantity of cells and their startling diversity can come away with the sense that each cell is easily expendable and of little consequence. But the same body that impresses us with specialization and diversity also affirms that *each* of its many members is valuable and often essential for survival.

Interestingly, the worth of each member is also the aspect most often stressed in biblical imagery of the Body of Christ (see Romans 12:5, 1 Corinthians 12, and Ephesians 4:16). Listen to the mischievous way in which Paul expresses himself in 1 Corinthians: "Those parts of the body that seem to be weaker are indispensable, and the parts that we think are less honourable we treat with special honour. And the parts that are unpresentable are treated with special modesty, while our presentable parts need no special treatment. But God has combined the members of the body and has given greater honour to the parts that lacked it, so that there should be no division in the body, but that its parts should have equal concern for each other. If one part suffers, every part suffers with it; if one part is honoured, every part rejoices with it". (vv. 22–26)

Paul's point is clear: Christ chose each member to make a unique contribution to His Body; Without that contribution, the Body could malfunction severely. Paul underscores that the less visible members (I think of organs like the pancreas, kidney, liver, and spleen) are perhaps the most valuable of all. Although I seldom feel consciously grateful for them, they perform daily functions that keep me alive.

I must keep coming back to the image of the body, because in our Western societies the world of persons is determined by how much society is willing to pay for their services. Aeroplane pilots, for example, must endure rigorous education and testing procedures before they can fly for commercial airlines. They are then rewarded with luxurious lifestyles and societal respect. Within the corporate world, visible symbols such as office furnishings, bonuses, and salaries announce the worth of any given employee. As a person climbs, he or she will collect a sequence of important-sounding titles (the U.S. government issues a book cataloguing ten thousand of them).

In the military, the chain of command defines a person's worth. One salutes superior officers, gives orders to those of lower rank, and one's uniform and stripes alert everyone to his or her relative status. In civil service status is reflected in an individual's "GS grade," a numerical label.

Our culture is shot through with rating systems, beginning from the first grades of school when children receive marks defining relative performance. That, combined with factors such as physical appearance, popularity, and athletic prowess, may well determine how valuable people perceive themselves to be.

Living in such a society, my vision gets clouded. I begin viewing janitors as having less human worth than jet pilots. When that happens, I must turn back to the lesson from the body, which Paul draws against just such a background of incurable competition and value ranking. In human society, a janitor has little status because he is so replaceable. Thus, we pay the janitor less and tend to look down on him. But the body's division of labour is not based on status; status is, in fact, immaterial to the task being performed. The body's janitors are indispensable. If you doubt that, talk with someone who must go in for kidney dialysis twice a week.

The Bible directs harsh words to those who show favouritism. James spelled out a situation we can all identify with: "Suppose a man comes into your meeting wearing a gold ring and fine clothes, and a poor man in shabby clothes also comes in. If you show special attention to the man wearing fine clothes and say, 'Here's a good seat for you,' but say to the poor man, 'You stand there,' or, 'Sit on the floor by my feet,' have you not discriminated among yourselves and become judges with evil thoughts?" He concludes, "If you show favouritism, you sin and are convicted by the law as lawbreakers. For whoever keeps the whole law and yet stumbles at just one point is guilty of breaking all of it". (James 2:2–4, 9–10)

Paul states the same truth positively, "Here there is no Greek or Jew, circumcised or uncircumcised, barbarian, Scythian, slave or free, but Christ is all, and is in all" (Colossians 3:11).

In our rating-conscious society that ranks everything from baseball teams to "the best chilli in New York," an attitude of relative worth can easily seep into the church of Christ. But the design of the group of people who follow Jesus should not resemble a military machine or a corporate structure. The church Jesus founded is more like a family in which the son retarded from birth has as much worth as his brother the Rhodes scholar. It is like the body, composed of cells most striking in their diversity but most effective in their mutuality.

God requires only one thing of His "cells": that each person be loyal to the Head. If each cell accepts the needs of the whole Body as the purpose of its life, then the Body will live in health. It is a brilliant stroke, the only pure egalitarianism I observe in all of society. He has endowed every person in the Body with the same capacity to respond to Him. In Christ's Body a teacher of three-year-olds has the same value as a bishop, and that teacher's work may be just as significant. A widow's dollar can equal a millionaire's annuity. Shyness, beauty, eloquence, race, sophistication – none of these matter, only loyalty to the Head, and through the Head to each other.

*

Our little church at Carville includes one devout Christian named Lou, a Hawaiian by birth, who is marked with visible deformities caused by leprosy. With eyebrows and eyelashes missing, his face has a naked, unbalanced appearance, and paralysed eyelids cause tears to overflow as if he is crying. He has become almost totally blind because of the failure of a few nerve cells on the surface of his eyes.

Lou struggles constantly with his growing sense of isolation from the world. His sense of touch has faded now, and that, combined with his near-blindness, makes him afraid and withdrawn. He most fears that his sense of hearing may also leave him, for Lou's main love in life is music. He can contribute only one "gift" to our church, other than his physical presence: singing hymns to God while he accompanies himself on an autoharp. Our therapists designed a glove that permits Lou to continue playing his autoharp without damaging his insensitive hand.

But here is the truth about the Body of Christ: not one person in Carville contributes more to the spiritual life of our church than Lou playing his autoharp. He has as much impact on us as does any member there by offering as praise to God the limited, frail tribute of his music. When Lou leaves, he will create a void in our church that no one else can fill – not even a professional harpist with nimble fingers and a degree from Julliard School of Music. Everyone in the church knows that Lou is a vital, contributing member, as important as any other member – and that is the secret of Christ's Body. If each of us can learn to glory in the fact that we matter little except in relation to the Body, and if each will acknowledge the worth in every other member, then perhaps the cells of Christ's Body will begin acting as He intended.

5

UNITY

We cannot live for ourselves alone. Our lives are connected by a thousand invisible threads, and along these sympathetic fibres, our actions run as causes and return to us as results.
HERMAN MELVILLE

The biologist takes from an incubator an egg containing a fully developed young chicken. Just fourteen days ago this egg was a single cell (the largest single cell in the world is an unfertilized ostrich egg). Now it is a mass of hundreds of millions of cells, a whirlpool of migrating protoplasm hurriedly dividing and arranging itself to prepare for life outside. The biologist cracks the shell and sacrifices the chick.

Though the embryo is now dead, some of its cells live on. Word travels fast through the body, but it may be days before the far outposts surrender. From the tiny heart the biologist extracts a few muscle cells and drops them in saline solution. Under the microscope the individual cells appear as long, spindly cylinders, crisscrossed like sections of railroad track. Their destiny is to throb, and they persist even in the anarchic world apart from the body. Each cell beats out an incessant rhythm – pitiful and useless palpitations when isolated from the chick. But if properly nourished, these lonely cells can be kept alive.

Unlinked by a pacemaker, the cells beat irregularly, spasmodically, each tapping out a rhythm approximate to the 350 beats a minute normal to a chick. But as the observer watches, over a period of hours an astonishing phenomenon occurs. Instead of five independent heart cells contracting at their own pace, first two, then three, and then all the cells pulse in unison. There are no longer five beats, but one. How is this sense of rhythm communicated in the saline, and why?

Some species of fireflies act similarly. A wanderer discovers a cluster of them in a jungle clearing, flickering haphazardly. As he watches, one by one the fireflies fall into synchronization until soon he sees not dozens of twinkling lights but one light, switched on and off, with fifty branch locations. The heart cells and the fireflies sense an innate rightness about playing the same note at the same time, even when no conductor is present.

Cooperation, a curious phenomenon of cells outside the body, is the essential regimen of life inside. There, every heart cell obeys in tempo or the animal dies. Each cell is flooded with communication about the rest of the body. How does the roaming white cell in the bat's wing know which cell to attack as invaders and which to welcome as friends? No one knows, but the body's cells have a nearly infallible sense of *belonging*.

All living matter is basically alike; a single atom differentiates animal blood from plant chlorophyll. Yet the body senses infinitesimal differences with an unfailing scent; it knows its hundred trillion cells by name. The first heart transplant recipients died, not because their new hearts failed, but because their bodies would not be fooled. Though the new heart cells looked in every respect like the old ones and beat at the correct rhythm, *they did not belong*. Nature's code of membership had been broken. The body screams

"Foreigner!" at imported cells and mobilizes to destroy them. This conundrum of the immune reaction keeps organ transplant science in its kindergarten phase.

To complicate the process of identity, the composite of Paul Brand today — bone cells, fat cells, blood cells, muscle cells — differs entirely from my components ten years ago. All cells have been replaced by new cells (except for nerve cells and brain cells, which are never replaced). Thus, my body is more like a fountain than a sculpture: maintaining its shape, but constantly being renewed. Somehow my body knows the new cells belong, and they are welcomed.

What moves cells to work together? What ushers in the higher specialized functions of movement, sight, and consciousness through the coordination of a hundred trillion cells?

The secret to membership lies locked away inside each cell nucleus, chemically coiled in a strand of DNA. Once the egg and sperm share their inheritance, the DNA chemical ladder splits down the centre of every gene much as teeth of a zipper pull apart. DNA re-forms itself each time the cell divides: 2, 4, 8, 16, 32 cells, each with the identical DNA. Along the way cells specialize, but each carries the entire instruction book of one hundred thousand genes. DNA is estimated to contain instructions that, if written out, would fill a thousand six-hundred-page books. A nerve cell may operate according to instructions from volume four and a kidney cell from volume twenty-five, but both carry the whole compendium.* It provides each cell's sealed credential of membership in the body. Every cell possesses a genetic code so complete that the entire body

---

* The DNA is so narrow and compacted that all the genes in all my body's cells would fit into an ice cube; yet if the DNA were unwound and joined together end to end, the strand could stretch from the earth to the sun and back more than four hundred times.

could be reassembled from information in any one of the body's cells, which forms the basis for speculation about cloning.

The Designer of DNA went on to challenge the human race to a new and higher purpose: membership in His own Body. And that membership begins with a stuff-exchange, analogous to an infusion of DNA, for each new cell in the Body. The community called Christ's Body differs from every other human group. Unlike a social or political body, membership in it entails something as radical as a new coded imprint inside each cell. In reality, I become genetically like Christ Himself because I belong to His Body.

The more I ponder the implications of this analogy, the more it illuminates for me a spiritual truth which the Bible states often but in puzzling terms:

"Do you realize that Christ Jesus is in you?" "I have been crucified with Christ and I no longer live, but Christ lives in me" – Paul. And, "I am in my Father, and you are in me, and I am in you." "I am the vine; you are the branches" – Jesus (2 Corinthians 13:5; Galatians 2:20; John 14:20; 15:5).

I can only fathom the concept of being visited by the living Christ by considering its parallel in the physical world: the mystery of life in which DNA passes on an infallible identity to each new cell. Christ has infused us with spiritual life that is just as real as natural life. I may sometimes doubt my new identity or perhaps *feel* like my old self, but the Bible statements are unequivocal. "Whoever believes in the Son has eternal life," said Jesus, "but whoever rejects the Son will not see life". (John 3:36) The difference between a person joined to Christ and one not joined to Him is as striking as the difference between a dead tissue and my organic body. DNA has organized chemicals and minerals to form a living, growing body, all of whose parts possess its unique corporate identity. In a parallel way, God uses the materials and genes of natural man, splitting them apart and recombining them with His own spiritual life.

Jesus made the interchange possible: the Virgin Birth assumes that His DNA was fully God and fully human joined in one. And now, through union with Him, I can carry within me the literal presence of God.

This unfathomable idea of an actual identity exchange is implicit in conversion. Jesus described the process in terms His hearers could understand. To Nicodemus He called it being "born again" or "born from above," indicating that spiritual life requires an identity change as drastic as a person's first entrance into the world.

As a result of this stuff-exchange, we carry within us not just the image of, or the philosophy of, or faith in, but the actual substance of God. One staggering consequence credits us with the spiritual genes of Christ: as we stand before God, we are judged on the basis of Christ's perfection, not our unworthiness. "If anyone is in Christ, he is a new creation; the old has gone, the new has come! . . . God made him who had no sin to be sin for us, so that in him we might become the righteousness of God (2 Corinthians 5:17, 21). Elsewhere, Paul underscored, "Your life is now hidden with Christ in God" (Colossians 3:3). We are "in Him" and He is "in us".

Just as the complete identity code of my body inheres in each individual cell, so also the reality of God permeates every cell in His Body, linking us members with a true, organic bond. I sense that bond when I meet strangers in India or Africa or California who share my loyalty to the Head; instantly we become brothers and sisters, fellow cells in Christ's Body. I share the ecstasy of community in a universal Body that includes every man and woman in whom God resides.

Along with the incredible benefits of our identity transfer come some sobering responsibilities. When we act in the world, we quite literally subject God to that activity. Paul applied the body analogy to impress upon promiscuous Corinthians the full extent of their new identity. "You are members of Christ's Body," he warned. "Shall I then take the members of Christ and unite them with a prostitute? Never! Do you not know that he who unites himself with a prostitute is one with her in body?" And, he concluded, "You are not your own; you were bought at a price. Therefore honour God with your body" (1 Corinthians 6:15–16, 19–20).

I cannot imagine a more sobering argument against sin. Paul appeals not to a guilt-inducing "God is watching you" argument, but to a mature realization that we literally incarnate God in the world. It is indeed a heavy burden.

The process of joining Christ's Body may at first seem like a renunciation. I no longer have full independence. Ironically, however, renouncing my old value system – in which I had to compete with other people on the basis of power, wealth, and talent – and committing myself to Christ, the Head, abruptly frees me. My sense of competition fades. No longer do I have to bristle against life, seizing ways to prove myself. In my new identity my ideal has become to live my life in such a way that people around me recognize Jesus Christ and His love, not my own set of distinctive qualities. My worth and acceptance are enveloped in Him. I have found this process of renunciation and commitment to be healthy, relaxing, and wholly good.

6
SERVICE

It is in giving that we receive, it is in pardoning that we are pardoned, it is in dying that we are born again to eternal life.

SAINT FRANCIS OF ASSISI

I close my eyes and reflect on my life, flashing back through my memories to recall rare moments of intense pleasure and fulfilment. To my surprise, my mind passes by hedonistic recollections of great meals, thrilling vacations, or awards ceremonies. Instead, it settles on instances when I have been able to work closely with a team and our work has allowed us to serve another human being. Sometimes our work has helped to improve sight, arrest crippling effects of leprosy, or save a leg from amputation. At the time, some of those acts involved apparent sacrifice. Surgeries were performed in primitive situations on a portable table, in 110-degree heat, with a young assistant beside me holding a flashlight. But those times of working together, when I focused all my concentration on the goal of helping another, glow with an unusual lustre.

I remember one patient particularly, Sadagopan, or Sadan as his friends called him. Born of an artistic, high-caste family in South India, he was educated and refined, but leprosy had made him an outcast. Passers-by on the street, noticing his sores, would call him names, shrinking from him in disgust. Cafés would not serve him; buses would not admit him.

Sadan came to our hospital at Vellore filled with despair. Though his face looked normal, his fingers were shortened and paralysed, and his ulcerated feet left damp spots on the floor wherever he stepped. Constant infection in the bones had reduced his feet to half their normal length. Sadan was well advanced in the classic leprosy process of losing all use of hands and feet, a process our medical team had been desperately fighting to reverse.

We were convinced that most foot destruction of this type resulted from the stress of walking on feet without sensation. Simple observation had indicated that; for we were able to match up nails and rough spots in patients' shoes with the ulcers on their feet. If only we could spread the stress evenly over the whole surface of the foot, perhaps the skin could bear it and our patients could walk without further damage.

Sadan was an ideal person to test our theory. He eagerly agreed to live in a mud-and-thatch hut in our New Life Centre and volunteered for anything that might improve his condition. We put him to bed until his ulcerous feet had healed and then fitted him with soft sandals. Excitedly, he began to walk. But in less than a week a runny red blister appeared on his foot, and Sadan returned to bed. We all remained cheerful, however, for the programme was still experimental; we merely needed to keep seeking the right footwear.

There is no way to compress into a few paragraphs our accumulating emotions of hope and despair that see-sawed through the next three years. We tried plaster casts, wooden clogs, and plastic shoes formed from wax moulds. I travelled to Calcutta to learn how to mix polyvinyl chloride and to England to test spray-on plastics.

I felt as though I was trying, and failing, to sustain the life of two cherished friends. One was the theory, a conviction born and nurtured in my own mind, that leprosy deformity could be prevented. The disease attacked nerves primarily, I believed, so we only had to find a way to protect patients from self-destruction. We had collected much supporting evidence and had achieved successes with less severe cases. More than a cold, scientific theory, the idea was almost like our own child. In the face of opposition from older, more experienced doctors, our little group at Vellore was fighting for a cause that could conceivably overturn ancient prejudice against leprosy. Now, over months and ultimately years, as Sadan tried shoe after shoe and we saw the ulcers recur and heal and recur again, our idea was dying.

But there was another friend to sustain: Sadan himself. After all, it was his feet we were studying. *We* gambled with ideas; Sadan offered his body and hope. I got to the point where I could hardly bear to meet him and remove his socks, though I knew I would never hear a complaint from him. I had come to love Sadan, and I knew he loved me and clung to me as his last hope. I often thought that for his sake I should give up and amputate his feet. At least then, with wooden legs, he could return to his home and family.

After each failure we would start with a new design – a high, firm boot or a flexible, springy sole – and then each evening we would meet with rising hope.

Sometimes a month would go by. ''Sadan, now we've really found it!'' I would exclaim when I could find no sign of infection. But finally, inevitably, the signal of failure appeared.

I would relieve pressure on an area where there had been an old problem only to have the foot rub in a new area.

Our team comforted Sadan; he comforted us. We all wept on our own and tried not to let our despair show.

Besides the shoe's design, I also had to contribute hard manual work. After a day of teaching and surgery I would make my way to the New Life Centre and revive my old trade of carpentry. With a set of chisels and gouges and rasps I would shape a block of wood into a clog, then thin it down to a copy of Sadan's foot. As he sat on the bench, I would match every bump on his foot with a hollow in the clog. Finally I would sandpaper the finish to a smooth, polished texture that could not damage his feet. After fastening leather straps, I would launch Sadan on another period of experimental walking.

During the next weeks I would check pressures and feel his feet for signs of inflammation, adjusting the clogs accordingly. Finally Sadan would bring me one of the clogs and point to a blood stain near one end. "I'm sorry," he would say. I, too, would mumble "I'm sorry," and we would start over again.

Amid all the despondency, though, there were some good moments. We learned that all the most successful shoes had a "rocker" — a rigid bar under the sole of the shoe that made the foot rock, like a see-saw on a pivot, instead of bend. Most importantly, I discovered I could feel signs of damage in the foot. Although Sadan could not sense pain, my hands could detect an area of heat in the tissues. I quickly learned that this indicated early damage, and in a day or two that spot would break down. By finding these spots early, I could alter the shoe or give the foot a rest so the skin could recover.

Soon after this discovery the periods of successful walking grew longer and the breakdowns were less frequent. An almost breathless hope began to replace our despair. Sadan went for months without trouble and was walking better than he had for years.

Then came an almost blinding insight. One day I was feeling his feet, which I had come to know better than my own, rejoicing that they were cool and free from inflammation. Suddenly I realized that his skin felt different. Sadan's skin had always seemed solid and warm and tense, whereas now it was loose and cool, almost shrunken. Then it hit me. These were Sadan's normal feet coming through for the first time. In all the years I had known him, a residue of chronic infection and recurrent damage had kept his feet swollen and inflamed. Now, after several damage-free months, the protein fluids were being absorbed, leaving skin and bone free from inflammation and thus able to move and adjust to outside pressures.

At least part of our earlier problem had stemmed from the fact that Sadan's feet had never returned to normal between his troubles. The very tissue his body had developed to fight infection also made him vulnerable to mechanical stress. We had him walking too soon after each apparent healing, but with dulled nerves, Sadan did not notice. Over the months I had learned to feel the pain he could not feel.

Today when I visit India I make a small detour to visit my dear friend Sadan, his wife Kokela, and their fine family. He is proud and independent, earning his living as a hospital record librarian. He walks on a type of rocker boot, now used in many parts of the world for leprosy patients, diabetics, and others with insensitive feet.

When we meet, Sadan always takes off his boots and enthusiastically displays his feet that have remained free of ulcers for many years. His skin is loose and free, and his feet are cool. I run my fingers over every familiar contour. When our eyes meet, we remember our days of despair and tears. But we remember most vividly the ecstasy of that day when we knew his real feet had at last won through. I call them my feet now, as he says my hands are his, because only through them can he feel.

*

When Jesus described the Christian life, often His invitation to it sounded more like a warning than a sales pitch. He spoke of "counting the cost," of selling all and "taking up a cross" to follow Him. While that attitude used to puzzle me greatly, I now believe He was simply underscoring the need for loyalty, which in biological terms means the need for individual cells to offer up service for the whole body. Sometimes following the Head may involve a sort of self-denial, including some pain. But I have learned, through rare instances like my experience with Sadan, that service also opens up levels of fulfilment far exceeding any others I have encountered. We are called to self-denial, not for its own sake, but for a compensation we can obtain in no other way.

Our culture exalts self-fulfilment, self-discovery, and autonomy. But according to Christ, it is only in losing my life that I will find it. Only by committing myself as a "living sacrifice" to the larger Body through loyalty to Him will I find my true reason for being.

We cling to a self-serving feeling of martyrdom about such a life of service. In actual fact, we are called to deny ourselves in order to open up to a more abundant life. In the exchange, the advantage clearly rests on our side: crusty selfishness peels away to reveal the love of God expressed through our own hands which, in turn, reshapes us into His image. "To refuse to deny one's self" said Henry Drummond, "is just to be left with the self undenied."

The concept of service is best communicated through a personal example rather than through an abstract discussion, and a powerful memory edges into my mind of a strange-looking Frenchman named Abbe Pierre. He arrived at the leprosy hospital at Vellore wearing his simple monk's habit and carrying a blanket over his shoulders and one carpetbag containing everything he possessed. I invited him to stay at my home, and there he told me his story.

As a Catholic friar he had been assigned to work among the beggars in Paris after World War II. At that time beggars in that city had nowhere to go, and in winter many of them would freeze to death in the streets. Abbe Pierre began by trying to interest the community in the beggars' plight, but was unsuccessful. He decided the only recourse was to show the beggars how to mobilize themselves. First, he taught them to do their tasks better. Instead of sporadically collecting bottles and rags, they organized into teams to scour the city. Next, he led them to build a warehouse from discarded bricks and start a business in which they sorted vast amounts of used bottles from big hotels and businesses. Finally Pierre inspired each beggar by giving him responsibility to help another beggar poorer than himself. Then the project really began to succeed. An organization called Emmaus was founded to perpetuate Pierre's work, with branches in other countries.

Now, he told me, after years of this work in Paris, there were no beggars left in that French city. Pierre believed his organization was about to face a serious crisis.

"I must find somebody for my beggars to help!" he declared and had begun searching in other places around the world. It was during one of those trips that he had come to Vellore. He concluded by describing his dilemma. "If I don't find people worse off than my beggars, this movement could turn inward. They'll become a powerful, rich organization and the whole spiritual impact will be lost! They'll have no one to serve." As we walked out of the house toward the student hostel to have lunch, my head was ringing with Abbe Pierre's earnest plea for "somebody for my beggars to help!"

We had a tradition among the medical students at Vellore about which I warned all guests in advance. All lunchtime guests would stand and say a few words about who they were and why they had come. Like students everywhere, ours were lighthearted and ornery, and they had developed an unspoken three-minute tolerance rule. If any guest talked longer than three minutes (or became boring before that time was up), the students would stamp their feet and make the person sit down.

On the day of Pierre's visit, he stood up and I introduced him to the group. I could see the Indian students eyeing him quizzically – this small man with a big nose and nothing attractive about him, wearing a peculiar old habit. Pierre started speaking in French, and a fellow worker named Heinz and I strained to translate what he was saying. Neither of us was well-practised in French, since no one in that part of India spoke it, so we could only break in here or there with a summary sentence.

Abbe Pierre began slowly but soon speeded up, like a tape recorder turning too fast, with sentences spilling over each other, gesticulating all the while. I was extremely tense because he was going into the whole history of the movement, and I knew the students would soon shout down this great, humble man. Worse, I was failing miserably to translate his rapid-fire sentences. He had just visited the UN headquarters where he had listened to dignitaries manipulate fine-sounding, flowery words to express insults to other countries. Pierre was saying that you don't need language to express love, only to express hate. The language of love is what you *do*. Then he spoke faster and faster, and Heinz and I looked at each other and shrugged helplessly.

Three minutes passed, and I stepped back and looked around the room. No one moved. The Indian students gazed at Pierre with piercing black eyes, their faces rapt. He went on and on, and no one interrupted. After twenty minutes Pierre sat down, and immediately the students burst into the most tremendous ovation I ever heard in that hall.

Completely mystified, I had to question some of the students. "How did you understand? No one here speaks French."

One student answered me, "We did not need a language. We felt the presence of God and the presence of love."

Abbe Pierre had learned the discipline of loyal service that determines the body's health. He had come to India and found leprosy patients to fulfil his desperate search to find someone worse off than his beggars, and when he found them, he was filled with love and joy. He returned to his beggars in France, and they and Emmaus worked to donate a ward at the hospital in Vellore. They had found people who needed their help so the spiritual motives of their lives continued on. The Emmaus movement thus flourished as a serving part of Christ's body.

7
MUTINY

> *An enormous technology seems to have set itself the task of*
> *making it unnecessary for one human being ever to ask anything*
> *of another in the course of going about his daily business.*
> *We seek more and more privacy, and feel more and more*
> *alienated and lonely when we get it.*
>
> PHILIP SLATER

At the central railway station in Madras, India, lay a beggar
woman more pitiful than the others I saw there. She had
positioned herself alongside the stream of passengers
hurrying to catch their trains. Businessmen with brief-cases
passed by her, as did wealthy tourists and government
officials.

Like many Indian beggars, the woman was emaciated,
with sunken cheeks and eyes and bony limbs. But,
paradoxically, a huge mass of plump skin, round and sleek
like a sausage, was growing from her side. It lay beside
her like a formless baby, connected to her by a broad bridge
of skin. The woman had exposed her flank with its
grotesque deformity to give her an advantage in the rivalry
for pity. Though I saw her only briefly, I felt sure that the
growth was a lipoma, a tumour of fat cells. It was a part
of her and yet not, as if some surgeon had carved a hunk
of fat out of a three hundred pound person, wrapped it in

live skin, and deftly sewed it on this woman. She was starving; she feebly held up a spidery hand for alms. But her tumour was thriving, nearly equalling the weight of the rest of her body. It gleamed in the sun, exuding health, sucking life from her.

*

Fat cells: the Madras beggar's tumour was composed entirely of an orgiastic community of them. In our figure-conscious Western culture, the word "fat" connotes a lack of discipline, an unnecessary aggregation of cells that should be reduced.

From the surgeon's vantage point, however, as he draws a knife across the skin, exposing oleaginous layers of fat cells, the evil connotation is balanced by a sense of the value of fat. It insulates against cold, and for that reason billions of fat cells congregate just below the skin. (Because of this, fat people can survive cold air and water better than thin people.) Fat cells pitch their tents wherever they find space around internal organs and muscles and between layers in the body. Their presence helps cushion those vital parts against jarring shocks.

Nothing influences appearance as much as fat. Why are young women so pleasing to the eye? An abundance of fat cells fills in the irregularities of bone and muscle, giving their skin a sleek, smooth contour.

But there is more to fat's function than insulation and contouring. Each fat cell is a storehouse containing a yellow globule of oil which crowds out the cell nucleus. Most of the time the cell lies dormant, while the body eats enough food to fuel its needs. Come famine, people with plenteous fat cells will be able to sit by while others starve. And that is the most strategic function of fat.

When all is going well, the body takes in just enough

food to maintain itself, grow, and replace worn cells. But when the supply diminishes, as when a person mowing the lawn delays supper in order to utilize the summer light, a signal sounds in the body's fat cells. To the liver short of glycogen and the blood short on glucose, the fat cells freely yield their oily treasure. By being the body's storehouse, the fat cells free other cells to do their job more efficiently. For example, if every muscle cell had to include a pouch-like reservoir of energy, our bodies would be deformed lumps and nodules.

Some fat is readily expendable: it goes first when a person starts a diet. Other fat, such as that around the kidney and in the palm of the hand, holds out becasue of its important secondary functions. When the body is starving, however, even these high priority fat cells must relinquish their important contents.

I like to think of fat cells as the banker cells of the body. In times of plenty they bulge with excess, as the body deposits more than it withdraws. In times of want they channel their chemical wealth back into the bloodstream.

Sometimes a dreaded thing occurs in the body – a mutiny – resulting in a tumour lipoma such as the one attached to the Madras beggar. A lipoma is a low-grade, benign tumour. It derives from a single fat cell, skilled in its lazy role of storing fat, that rebels against the leadership of the body and refuses to give up its reserves. It accepts deposits but ignores withdrawal slips. As that cell multiplies, daughter cells follow its lead and a tumour grows like a fungus, filling in crevices, pressing against muscles and organs. Occasionally a lipoma crowds a vital organ like the eye, pushing it out of alignment or pinching a sensitive nerve, and surgery is required.

I have removed such lipoma tumours. Under a

microscope they seem composed of healthy fat cells, bulging with shiny oils. The cells function beautifully except for one flaw – they have become disloyal. In their activity they disregard the body's needs. And so the beggar woman in Madras gradually starved while a lipoma that was part of her engorged itself.

A tumour is called benign if its effect is fairly localized and it stays within membrane boundaries. But the most traumatizing condition in the body occurs when disloyal cells defy inhibition. They multiply without any checks on growth, spreading rapidly throughout the body, choking out normal cells. White cells, armed against foreign invaders, will not attack the body's own mutinous cells. Physicians fear no other malfunction more deeply: it is called cancer. For still mysterious reasons, these cells – and they may be cells from the brain, liver, kidney, bone, blood, skin, or other tissues – grow wild, out of control. Each is a healthy, functioning cell, but disloyal, no longer acting in regard for the rest of the body.

Even the white cells, the dependable palace guard, can destroy the body through rebellion. Sometimes they recklessly reproduce, clogging the bloodstream, overloading the lymph system, strangling the body's normal functions – such as leukaemia.

*

Because I am a surgeon and not a prophet, I tremble to make the analogy between cancer in the physical body and mutiny in the spiritual body of Christ. But I must. In His warnings to the church, Jesus Christ showed no concern about the shocks and bruises His Body would meet from external forces. "The gates of hell shall not prevail against my church," He said flatly. (Matthew 16:18) He moved easily, unthreatened, among sinners and criminals. But He cried out against the kind of disloyalty that comes from within.

I must concentrate on how I, as an individual cell, should respond to the crying needs of the Body of Christ in other parts of the world. Beyond that, I cannot and should not make sweeping judgments about what the response of other Christians should be.

But I must say, from the perspective of a missionary who spent eighteen years in one of the poorest countries on earth, the contrasts in resources are astonishingly large. At Vellore we treated leprosy patients on three dollars per patient per year; yet we turned many away for lack of funds. Then we came to America where some churches were heatedly discussing their million-dollar gymnasiums and the cost of landscaping and fertilizer and a new steeple . . . and sponsoring seminars on tax shelters for members to conserve their accumulated wealth. As I saw those churches' budgets for foreign missions and for inner city work, I could not force a telling image from my mind – the memory of the Madras woman slowly starving to death while her lipoma grew plump and round.

The problem is not just an American problem, or even a Western problem. I could easily point to examples of hoarding in every society I've seen: in the cruel Iks of Africa, in Soviet Russia, in the disparity within the Christian community in India. The warning applies to all of us. My only message is the caution of a doctor: remember, the body will have health only if each cell regards the needs of the whole body.

I wonder if perhaps we in the West get caught up in a competitive spiral with "cells" around us and become oblivious to the stark needs of the rest of the world. In the Body of Christ ownership of property and money is no sin; it is an important function of certain members.

And when I liken wealthy people to fat cells, I use the image positively, as an admiring doctor who appreciates the role of fat. Hospitality and generosity are made easier by wealth. Reserves can help the Body care for itself and fuel its muscular activity in a hurting world. However, the control of fat is a difficult problem, both in biology and in religion.

I will cite two sets of statistics and then leave the application to you.

First, wealth is not only material in nature. Ninety per cent of all the full-time Christian workers in the world work in North America among less than 10 per cent of the world's population. On Sunday morning in rural Louisiana I can flip a radio dial to ten different church services, while elsewhere entire countries have no Christian witness.

Second, consider the world as if it were shrunk down to a community of 1000 persons:

In our town of 1000 –

180 of us live high on a hill called the developed world;
820 live on the rocky bottom land called the rest of the world.

The fortunate 180 on the hill have 80 per cent of the wealth of the whole town, over half of all the rooms in town with over two rooms per person, 85 per cent of all the automobiles, 80 per cent of all the TV sets, 93 per cent of all the telephones, and an average income of $5000 per person per year.

The not-so-fortunate 820 people on the bottom get by on only $700 per person per year, many of them on less than $75. They average five persons to a room.

How does the fortunate group of hill-dwellers use its incredible wealth? Well, as a group they spend less than

1 per cent of their income to aid the lower land. (In the United States, for example, of every $100 earned:

$18.30 goes for food
$6.60 is spent on recreation and amusement
$5.80 buys clothes
$2.40 buys alcohol
$1.50 buys tobacco
$1.30 is given for religious and charitable uses, and only a small part of that goes outside the U.S.[1])

I wonder how the villagers on the crowded plain – a third of whose people are suffering from malnutrition – feel about the folks on the hill?

I realize these issues have complex economic and cultural factors behind them. But I am impressed with how decisively the early church responded to pressing needs: the apostle Paul took months out of his schedule to collect money from Greek Christians to aid improverished Jewish Christians in Jerusalem.

We need to pause and look carefully at ourselves. God needs all types of cells in His Body: fat and thin, rich and poor, simple and complex. But He only needs loyal cells. And in the area of using resources, Jesus, our Head, had many unsettling things to say. God save us from being a cancer within His Body.

BONES

8

A FRAME

Bone is power. It is bone to which the soft parts cling, from which they are, helpless, strung and held aloft to the sun, lest man be but another slithering earth-noser.

RICHARD SELZER

The setting was worthy of a horror movie. Each morning I threaded through dark, narrow corridors until I came to a winding stairway which led to the ancient attic. There I found rows of boxes, layered with dust, containing six hundred skeletons. Each day I crouched down over the boxes in the dimly lighted, creaking room, sorting through bones. In all, I spent seven days crouched in the musty attic of the old house in Copenhagen.

The house served as a museum for Dr Möller Christiansen, a medical historian, who had invited me there because the six hundred skeletons once belonged to people with leprosy. After discovering the bones on an island off the coast of Denmark and diligently studying them, Dr Christiansen had written an extraordinary book on leprosy. Those of us who worked with the disease could hardly believe it when we learned he had never observed a living patient. All his insights into the disease were drawn from the five hundred-year-old skeletons in his attic; yet he had taught us many facts about leprosy and had made good suggestions about its treatments.

Picking over his clattery bones, much as a child rummages through a box of precious toys, Dr Christiansen would locate certain favourites and proudly show me their features. Many skeletons, for example, had loose or missing front teeth, caused by leprosy's tendency to attack first the body's cooler parts. Together we examined bones of feet and hands, speculating what injuries could have caused their deformities.

One morning, working alone in the attic, I came across some boxes of skeletons that had been dug up from a monastery. I was soon to be reminded of a lecture given by anthropologist Margaret Mead, who spent much of her life studying primitive cultures. She asked the question, "What is the earliest sign of civilization?" A clay pot? Iron? Tools? Agriculture? No, she claimed. To her, evidence of the earliest true civilization was a healed femur, a leg bone, which she held up before us in the lecture hall. She explained that such healings were never found in the remains of competitive, savage societies. There, clues of violence abounded: temples pierced by arrows, skulls crushed by clubs. But the healed femur showed that someone must have cared for the injured person – hunted on his behalf, brought him food, and served him at personal sacrifice. Savage societies could not afford such pity. I found similar evidence of healing in the bones from the churchyard. I later learned from Dr Christiansen that an order of monks had worked among the victims: their concern came to light five hundred years later in the thin lines of healing where infected bone had cracked apart or eroded and then grown back together.

After a week I left that eerie attic feeling as though I had watched a slide show on an ancient civilization. My clues for visualizing it had been tiny projections and furrows on the surfaces of bones exhumed from the dust of history, but they had taught me much. Faces, hair, and clothes, which consume so much cultural energy, had all rotted away, leaving bones as the only mementoes of that settlement.

The bulky pelvis, for instance, quickly betrayed the sex of the person who owned it. A broad and shallow one with a smooth inner ring obviously belonged to a woman. The oval opening precisely matched the size and shape that a baby's head would need to squeeze through. The pelvis beside it, a man's, was narrower, more heart-shaped, and formed with heavier bones. Hard, nub-like projections on its inner ring marked where muscles and ligaments had once been attached.*

A closer look at bones such as those in Copenhagen shows surfaces which are not even and shiny, but coarsely filled with grooves for blood vessels and slick areas for gliding tendons. The very thickness of a bone may divulge its former use. Discus throwers and weight lifters have the densest bones because an exercised bone collects more calcium for needed strength. By carefully studying the stress lines of individual bones under a microscope, even a person's occupation can be guessed at. A horseback rider leaves definite clues in leg bones and pelvis. A porter who lugs heavy suitcases in his right hand will bear the effects of this stress in hip and shoulder.

Shakespeare said, "The good [men do] is oft interred with their bones." More than good is interred there. A field

* The female runner still lags behind the male, and blame rests on the pelvis. The projections on the man's pelvis allow for more powerful muscles, but a woman equipped with them could not bear a child. Similarly, a man's hip sockets are closer together, nearer the centre of gravity, which enables more efficient movement. If a woman's were similarly designed, there would be no room for the baby's head to extrude. So the odd pelvic bone represents a summation of many different requirements. When a woman wishes she could run faster or sway less or have a narrower base, let her know that the survival of the human race depends upon her being just the shape she is.

of science, forensics, exists to unravel the clues hidden in bones. Experts can determine a skeleton's age by how hard or "ossified" the cartilage has become. By age fifteen, for example, the foot is fully formed, at twenty-five the collarbone is fused to the breastbone, and by age forty, three-fourths of the seams in the skull have coalesced.

*

Simple laboratory experiments reveal the components of bones. Burning a bone in a fire will flare away all organic material, leaving an object the same shape and appearance as the bone but consisting of minerals only. If baked long enough, the bone will crumble between the fingers.

Hydrochloric acid does the opposite: it dissolves all minerals, leaving the organic substance collagen, again in the original shape. The treated object looks the same but is no longer bone. It has lost its hardness and cannot support weight. Such a bone can be tied into a knot and, when untied, will spring back into its shape. (Collagen makes even untreated bone surprisingly elastic: Arab children play with bows made from the ribs of camels.) Grit and glue – those are the ingredients of bone. We need both.

No Exxon researcher has yet discovered a material as well-suited for the body's needs as bone, which comprises only one-fifth of our body weight. In 1867 an engineer demonstrated that the arrangement of bone cells forms the lightest structure, made of least material, to support the body's weight. No one has successfully challenged his findings. As the only hard material in the body, bone possesses incredible strength, enough to protect and support every other cell. Sometimes we press our bones together like a steel spring, as when a pole vaulter lands. Other times we nearly pull a bone apart, as when my arm lifts a heavy suitcase.

In comparison, wood can withstand even less pulling tension, and could not possibly bear the compression forces that bone can. A wooden pole for the vaulter would quickly snap. Steel, which can absorb both forces well, is three times the weight of bone and would burden us down.

The economical body takes this stress-bearing bone and hollows it out, using a weight-saving architectural principle it took people millennia to discover; it then fills the vacant space in the centre with an efficient red blood cell factory that turns out a trillion new cells per day. Bone sheathes life.

I find bone's design most impressive in the tiny, jewellike chips of ivory in the foot. Twenty-six bones line up in each foot, about the same number as in each hand. Even when a soccer player subjects these small bones to a cumulative force of over one thousand tons per foot over the course of a match, his living bones endure the violent stress, maintaining their elasticity. Not all of us leap and kick, but we do walk some sixty-five thousand miles, or more than two and one half times around the world, in a lifetime. Our body weight is evenly spread out through architecturally perfect arches which serve as springs, and the bending of knees and ankles absorbs stress. Unfortunately, we coax our feet to assume the shape of footwear fashion, sometimes tilting our heels high and cancelling all the effects of that balanced design.

Bone's strength is quiet, dependable. It serves us well, without fanfare, and comes to our attention only when we encounter a rude, fracturing stress that exceeds its own high tolerance.

*

In order to appreciate the invisible frame each of us wears inside, we must pause to consider the progression of skeletons in nature, which offers abundant variations. Much of the earth's hard surface, sedimentary rock, was left to us by microscopic creatures who died and cemented together, pooling their skeletons to form rock. Of these simple creatures, perhaps the most exquisite are the saltwater *Protozoa* called *Radiolaria*.

Recall the most perfect snowflake you have ever seen: a large, unblemished one that floats like a feather on a frosty day. It has only six sides, but an abundance of symmetrical cusps gives it beauty. Now imagine a three-dimensional snowflake with hundreds of crystalline shapes sprouting from its centre. Such is a *Radiolaria* skeleton, billions of which float through our oceans.

The ocean is a hungry place, and skeletons there are as likely to be required for protection as for movement. So for the *Radiolaria*, molluscs, scallops, nautiluses, crabs, lobsters, and starfish a skeleton becomes a place of refuge.

On land, however, dominated by the incessant tug of gravity, movement is everything. The fastest rabbit evades the coyote and the fleetest African cat dines on gazelle. Several million species mimic their oceanic counterparts with external skeletons, notably the vast insect world. But these can only grow so large or the burden of armour becomes insupportable. The largest insects, with their excreted exoskeletons, barely approach the size of the smallest birds or mammals.

Thus, we're back to the old distinction: higher and lower. The highest animals, called vertebrates, rule even in the ocean. An internal, *living* skeleton allows revolutionary advances. No longer need an animal outgrow his home and risk a vulnerable moulting period. Rather, the skeleton grows with the animal, and because of the hundreds of muscles attached to internal scaffolding, heretofore unheard-of feats can be performed.

Insects and spiders can run, jump, and fly, but only with an internal skeleton can an animal as large as a barn swallow flaunt gravity with his skydiving, or a creature like a condor support a ten-foot wingspan and soar on thermals for hours. Only with an internal skeleton can an elephant charge like thunder across the grasslands or an elk hoist his rack of antlers proudly towards the sky. Without bones, locomotion tends to revert to the most primitive: the segmented scrunching of an earthworm or the lubricated slide of a slug.

Bones do not burden us; they free us.

9

HARDNESS

There are an infinity of angles at which one falls, only one at which one stands.

G. K. CHESTERTON

People are never born without bones, but some are born with defective bones in a condition called brittle bone disease. When this occurs, the victim's bone consists of deposits of calcium without the organic material welding them together – the grit without the glue. A foetus with this disorder may survive the pressures of birth, but with half its bones broken. Just diapering such a child may break his or her fragile legs; a fall could break dozens of bones.

At our Carville hospital a patient was given massive doses of steroids during treatment of her leprosy, and as a result her bones became soft. She could fracture her foot by walking too briskly. Whenever I examined the woman and checked her X-rays for fractures, I was reminded that the most important feature of bone is its hardness. That one property separates it from all other tissue in the body, and without hardness bone is virtually useless.

An analogous body as advanced and active as the Body of Christ's followers also needs a framework of hardness to give it shape, and I see the church's doctrine as being just such a skeleton. Inside the Body lives a core of truth that never changes – the laws governing our relationships to God and to other people.

Do I hear a groan? Our age smiles kindly on musings about unity and diversity and the contributions of individual cells. But the drive which stirred church councils and framers of the Constitution has stalled. Bones are dusty, crumbling, dead, belonging in a musty museum display case. Other parts of the body are memorialized: the heart on Valentine's Day, the sexual parts and the muscles in magazines and fashion, the hands in sculptures. The skeleton is relegated to Hallowe'en, a spooky remnant of the past, leeringly inhuman.

Today one can easily muster up sympathy and support for Jesus' ethics governing human behaviour. But squeezed in between His statements on love and neighbourliness are scores of harsh, uncompromising statements about our duties and responsibilities and about heaven and hell.

The modern world is still pictured as a courtroom scene, as described by the ancients, but not with God as Judge, setting the rules and arbitrating disputes. Rather, He stands indicted, and prosecutors are stalking across the stage jabbing their fingers at Him, demanding to know why He allows such a miserable world to continue and what right He has to make such grandiose claims about His Son. Don't all religions ultimately point to God? Isn't belief really an individual quest for ultimate meaning that each must find in his or her own way? What is this talk about "No man comes to the Father except by me" and "I am the way, the truth and the life"?

As I encounter the Body of Christ, I keep hitting against the hard tissue, the principles that do not change. Joining that Body involves a capitulation which defies my nature, an acknowledgment that someone else, not I, has already determined the way I should live. In some areas of my life I gladly accept restrictive laws. For instance, traffic laws inhibit my freedom (what if I don't want to stop?), yet I accept the inconvenience. I assume some skilled engineer calculated the number of one-way streets and red lights, and even if I doubt his ability, I prefer traffic laws to auto anarchy. But something within me rebels against being told how to live morally.

I came across this property of hardness when I was first taught about God. God is perfect, I was told, and cannot tolerate sin. His character requires Him to destroy sin whenever it is present, so I am branded an enemy of God. That fact, rooted in the first chapters of Genesis, is stressed throughout the Bible. God cannot ignore rebellion; His nature demands that justice be done, and nothing I can do will soften the inflexibility of that fact. I must meet Him on His terms, not my own.

Later, I learned how justice was accomplished. God obtained it on our behalf by becoming man and taking on Himself all the sin and rebellion we had stored up against Him. The debt of mankind was paid, but in a way that cost only God, not those of us who had piled it up. To the servant with a three-million-dollar debt Jesus announced, "It's forgiven; you owe nothing." And His message to the Prodigal Son: "The table is set; come join the party. The past may all be forgiven; all that counts is how you respond to what God has offered."

Even at its core, the hard, unchangeable part that does not flex, the gospel sounds almost like a fairy tale. "It's too good to be true," someone protested to Geroge MacDonald. "No," he replied, "it's so good it has to be true." The way back to God is hard, but only because there is just one way.

*

Others more skilled in theology than I must describe and interpret specific doctrines for us. Today some within the church attack law and doctrine. Situation ethics suggest that right and wrong often depend on the need and mood of the moment. I merely submit this single aspect of God's law: it must be consistent, like bone. Trust demands it.

I think back to an encounter with trust I had many years ago. Before I trained for surgery, I worked in the general practice of my father-in-law, near London. One day a woman came in with a list of complaints that exactly described gastritis. After a brief examination I told her my diagnosis, but she looked up at me with large, fear-filled eyes.

I repeated to her soothingly, "Really, it's not a serious condition. Millions of people have it, and with medication and care, you'll be fine." The fear did not fade from her face. Lines of tension were jerking in her forehead and jaw. To my "You'll be fine," she flinched as if I had said, "Your disease is terminal."

She quizzed me on every point, and I assured her I would be doing further tests to verify my diagnosis. She repeated to me all her symptoms and kept asking, "Are you sure? Are you sure?" So I ordered a barium meal and extensive X-rays.

When the test results came back, all pointed conclusively to gastritis. I saw the woman on one last visit. She trembled slightly as I spoke to her, and I used my most comforting and authoritative doctor's tone. "It is perfectly clear – no doubt – that you have gastritis. I thought so from the first visit, and now these tests have confirmed it. The condition is chronic and will require you to change diet and medication, but it should settle down. There is absolutely no reason for alarm."

The woman stared into my eyes with a piercing gaze for

at least a minute, as if she was trying to see into my soul.
I managed to hold her gaze, fearing that if I looked away
she would doubt me. Finally, she sighed deeply, and for
the first time her face relaxed. She sucked her breath in
sharply and said, "Well, thank you. I was sure I had cancer.
I had to hear the diagnosis from somebody I could trust,
and I think I can trust you."

She then told me a story about her mother, who had
suffered a long, painful disease. "One torturous night the
family doctor made a house call while mother was groaning
and pressing her hands to her stomach. She was feverish
and obviously suffering. When the doctor arrived, mother
said, 'Doctor, am I really going to get better? I feel so
ill and have lost so much weight . . . I think I must be
dying.'

"The doctor put his hand on my mother's shoulder,
looked at her with a tender expression, and replied, 'I know
how you feel. It hurts badly, doesn't it? But we can lick
this one – it is simply gastritis. If you take this medicine
for a little while longer, with these tranquillizers, we will
have you on your feet in no time. You'll feel better before
you know it. Don't worry. Just trust me.' My mother
smiled and thanked him. I was overwhelmed by the doctor's
kindness.

"In the hallway, out of her hearing, the doctor turned
to me and said gravely, 'I'm afraid your mother will not
last more than a day or two. She has an advanced case of
cancer of the stomach. If we keep her tranquillized, she
will probably pass away peacefully. If there's anyone you
should notify—'

"I interrupted him in mid-sentence. 'But, doctor! You
told her she was doing fine!'

" 'Oh, yes, it's much better that way,' he replied. 'She
does not know, and so she won't worry. She'll probably

die in her sleep.' He was right. My mother died that same night.''

This woman, now a middle-aged patient herself, had first gone to that same family doctor with her stomach pains. He had put a hand on her shoulder and said gently, "Don't worry. It's only gastritis. Just take this medicine, and you'll be feeling fine very soon." And he smiled the same paternal smile he had shown her mother. She had fled from his office in tears and would never see him again.

*

When people complain to me about the rigid, unbending laws of God, I think of that woman. The family doctor had obliterated all possibility of helping her because of his flexible attitude to truth. Only one thing could relieve her anxiety and despair: trust in someone who believed in truth that could not be twisted and bent.

Occasions will come when to be untruthful is more convenient or less offensive. But a respect for truth cannot be worn and then casually removed like a jacket; it cannot be contracted and then relaxed like a muscle. Either it is rigid and dependable, like healthy bone, or it is useless.

10
FREEDOM

He came to me as a patient in England: a brawny, burly Welshman who spoke lyrically and with a workman's vocabulary. "Mornin', doctor," he growled. As he removed his wool plaid jacket, I saw the reason for his coming. The upper part of his right arm was not pink skin, but grimy steel and leather – an awkward, brace-like contraption covered with black coal dust. I removed the brace. This was no artificial limb; his forearm was intact, but the flesh between his elbow and shoulder was flaccid. A long section of bone appeared to be missing. But if a mining accident had crushed his upper arm, how had his forearm survived?

After I studied the miner's records and X-rayed his arm, the puzzle fell into place. Years before, a bone tumour in his upper arm had led to a serious fracture which splintered large sections of bone there. Under the bright lights of the operating room his doctor had deftly stolen an eight-inch pipe of living bone and sewed back the muscles and skin around the space. As the miner lay recovering, his boneless arm seemed perfectly normal. Who would know the interior landscape had changed?

Everyone would know the first moment this miner used the muscles, still strong and intact, in his upper arm. Bones and muscles work on a triangle principle: the joint provides the fulcrum, and two bones work with a muscle. To pull the hand up, the biceps muscle, attached to the upper arm, pulls on the forearm. The arm bends at the elbow, and the triangle is complete. But one muscle and one forearm bone do not make a triangle; this coal miner lacked the third element, the bone of the upper arm.

Ever since his surgery years before, whenever the miner contracted his biceps muscle his entire upper arm shortened, like an earthworm spastically pulling in towards its middle. The fixed, resistant bone between his elbow and shoulder had become a soft, collapsible space, cancelling out the triangle that should have transferred force to his forearm. His ingenious Welsh doctor had fitted the miner with a crude exoskeleton, a bulky contraption of leather and steel which positioned stiff rods between his elbow and shoulder. When his biceps contracted, because these steel rods prevented his upper arm from merely shortening, the forearm could pull upward. The steel frame outside his arm worked much the same as the now-missing bone had inside his arm.

I have surgically removed such upper arm bones, though today we circumvent the awkwardness of an external skeleton by jamming a bone graft into the vacant space. A bone graft unites with the stumps above and below it, and gradually the arm will adjust to its new member. But this man's crude external brace had served him well for years, allowing him to work as a vigorous coal miner. He came to me asking for a new bone mainly because he was tired of having to buckle on his exoskeleton every day.

Because it is hard and sometimes subject to fracture, bone has acquired the reputation of a nuisance to human activity.

Bone prohibits us from squeezing into small spaces and from sleeping comfortably on hard ground. And what prevents skiers from adding twenty metres on to the looping, graceful ski jump and what keeps the slalom course in the domain of a few experts? The old nemesis of broken bones. A person who breaks a leg skiing could wish for stronger bones. But stronger bones would be thicker and heavier, making skiing far more limited or impossible.

No, the 206 lengths of calcium our body is strapped to are not there to restrict us; they free us. In the same way that the Welsh miner's arm was able to move only when it contained a proper scaffolding, external or internal, almost all our movements are made possible because of bone – rigid, inflexible bone.

*

In the Body of Christ also the quality of hardness is not designed to burden us; rather, it should free us. Rules governing behaviour work because, like bones, they are hard.

Moral law. The Ten Commandments. Obedience. Doing right. A "thou shalt not" negativism taints the words, and we tend to view them as opposites to freedom. As a young Christian, I cringed at such words. But later, especially after I became a father, I started thinking beyond my reflex reaction to the very nature of law. Are not laws essentially a description of reality by the One who created it? His rules governing human behaviour – are they not guidelines meant to enable us to live the very best, most fulfilling life on earth?

I do not slip easily into such reasoning. Laws are too encrusted with cultural barnacles that obscure their true essence. They can summon up in me deeply embedded memories of parental disapproval, and instead I crave another kind of freedom – freedom from law, not freedom by it.

I have discovered, however, that it is possible to see beyond the surface negativism of, for example, the Ten Commandments and to learn something of the true nature of laws. Rules soon seem as liberating in social activity as bones are in physical activity.

The first four of the Ten Commandments are rules governing a person's relationship to God Himself: Have no other gods before Me. Don't worship idols. Don't misuse My name. Remember the day set aside to worship Me. As I contemplate these once-forbidding commandments, more and more they sound like positive affirmations.

What if God had stated the same principles this way:

I love you so much that I will give you *Myself*. I am true reality, the only God you will ever need. In Me alone will you find wholeness.

I desire a wonderful thing: a direct, personal relationship between Myself and each of you. You don't need inferior representations of Me, such as dead wooden idols. You can have Me. Value that.

I love you so much that I have given you My name. You will be known as "God's people" on the earth. Value the privilege; don't misuse it by profaning your new name or by not living up to it.

I have given you a beautiful world to work in, play in and enjoy. In your involvement, though, set aside a day to remember where the world came from. Your bodies need the rest; your spirits need the reminder.

The next six commandments govern personal relationships. The first is already stated positively: honour your father and mother, a command echoed by virtually every society on earth. The next five:

Human life is sacred. I gave it, and it has enormous worth. Cling to it. Respect it; it is the image of God. He who ignores this and commits the sacrilege of murder must be punished.

The deepest human relationship possible is marriage. I created it to solve the essential loneliness in the heart of every person. To spread what is meant for marriage alone among a variety of people will devalue and destroy that relationship. Save sex and intimacy for its rightful place within marriage.

I am entrusting you with property. You can own things, and you should use them responsibly. Ownership is a great privilege. For it to work, you must respect everyone else's right to ownership; stealing violates that right.

I am a God of truth. Relationships only succeed when they are governed by truth. A lie destroys contracts, promises, trust. You are worthy of trust: express it by not lying.

I have given you good things to enjoy: oxen, grains, gold, furniture, musical instruments. But people are always more important than things. Love people; use things. Do not use people for your love of things.

Stripped down, the commandments emerge as a basic skeleton of trust that links relationships between people and between people and God. God claims, as the Good Shepherd, that He has given law as the way to the best life. Our own rebellion, from the Garden of Eden onward, tempts us to believe He is the bad shepherd whose laws keep us from something good.

Yes, one might reply, the Ten Commandments can be twisted around to reveal a more positive side. But why didn't God state them that way? Why did He say, "You shall *not* murder. You shall *not* commit adultery. You shall *not* steal. . . ."?

I suggest two answers. First, a negative command is actually less limiting than a positive one. "You may eat from any tree of the garden except one" allows more freedom than "You must eat from every tree of the garden, starting with the one in the northwest corner and working along the outer edge of the orchard." "You shall not commit adultery" is more freeing than "You must have sex with your spouse twice a week between the hours of nine and eleven in the evening." "Do not covet" is more freeing than "I am hereby prescribing limits on ownership. Every man is entitled to one cow, one ox, three gold rings. . . ."

Second, people were not yet ready for an emphasis on the positive commands. The Ten Commandments represent a kindergarten phase of morality: the basic laws needed for a society to operate. When Jesus came to earth, He filled in the positive side. Quoting the Old Testament, He summarized the entire law in two positive commands: "Love God with all your heart and with all your soul and with all your strength and with all your mind," and "Love your neighbour as you love yourself" (Luke 10:27). It is one thing not to covet my neighbour's property and not to steal from him. It is quite another to love him so that I care for his family as much as I care for mine. Morality took a quantum leap from prohibition to love. (Paul affirmed and developed this thought in Romans 13:8–10.)

Jesus' Sermon on the Mount puts the capstone on His attitude toward the law. There, He described the Ten Commandments as the bare minimum. They actually point to profound principles: modesty, respect, non-violence, sharing. Then Jesus submitted the ideal social ethic – a system governed by only one law, the law of love. He calls us toward that ideal. Why? So God can take a fatherly pride in how well His little experiment on earth is progressing? Of course not. These laws were not given for God's sake, but for ours. "The Sabbath was made for man, not man for the Sabbath," He said, and "You will know the truth, and the truth *will set you free*" (Mark 2:27; John 8:32). Jesus came to cleanse the violence, greed, lust, and hurtful competition from within us *for our sakes*. His desire is to have us become like God.

The Ten Commandments were the foetal development of bone, the first ossification from cartilage. The law of love is the fully developed, firm, liberating skeleton. It allows smooth movement within the Body of Christ, for it is hinged and jointed in the right places.

If you examine one law, like a random bone plucked from a pile, it may seem strangely shaped and illogical because laws, like bones, are designed for the complex, connected needs of a whole body. For example, as we have observed, the pelvis is a crazily shaped structure. It represents a compromise of converging needs: to walk, to protect abdominal organs, to sit, to support the back, and, in the woman's case, to bear children. Its shape exists to serve the body, not to dominate it. Similarly, God's laws governing us are a combination of conflicting human desires and needs, chosen to allow us to live life most fully and healthily. God, knowing our weaknesses and human frailties, designed the dogma of our faith and His laws to give strength and stability where we need them.

The law requiring sexual faithfulness in marriage to many people appears oddly and needlessly restrictive. Why not allow interchangeability, with men and women enjoying each other freely? We have the biological equipment for such practices. But sex transcends biology; it intertwines with romantic love, need for stable families, and many other factors. If we break one law, gaining the freedom of sexual experimentation, we lose the long-term benefits of intimacy that marriage is intended to provide. As my Welsh miner proved, removing one bone can ruin complex motion.

I have known people who feel compelled to cast off every possible limitation. They are like spoiled children, dashing from one toy to another, searching bitterly for an even better thrill, unaware that their search is actually a flight. Where do they stop cheating on their income tax? At what point do they allow the truth to break open before a cuckolded spouse? At what lie will their children cease to believe anything they say? Their lives become an entangling web of deception and fear. Does such a person have freedom?

I conclude with G. K. Chesterton that "the more I considered Christianity, the more I found that while it had established a rule and order, the chief aim of that order was to give room for good things to run wild."[1] He used the example of sex: "I could never mix in the common murmur of that rising generation against monogamy, because no restriction on sex seemed so odd and unexpected as sex itself. . . . Keeping to one woman is a small price for so much as seeing one woman. To complain that I could only be married once was like complaining that I had only been born once. It was incommensurate with the terrible excitement of which one was talking. It showed, not an exaggerated sensibility to sex, but a curious insensibility to it. A man is a fool who complains that he cannot enter Eden by five gates at once. Polygamy is a lack of the realization of sex; it is like a man plucking five pears in mere absence of mind."[2]

*

A skeleton is never beautiful; its contributions are strength and function. I do not inspect my tibia and wish it to be longer or shorter or more jointed. I just gratefully use it for walking, thinking about where I want to go rather than worrying about whether my legs will bear my weight. I should respond that way to the basic fundamentals of the Christian faith and the laws governing human nature. They are merely the framework for relationships which work best when founded on set, predictable principles. Of course, we can break them: adultery, thievery, lying, idolatry, oppression of the poor have crept into every society in history. But the result is a fracture that can immobilize the entire body. Bones, intended to liberate us, only enslave us when broken.

11
GROWTH

Better a little faith, dearly won, better launched alone on the infinite bewilderment of truth, than perish on the splendid plenty of the richest creeds.

HENRY DRUMMOND

In rural India legs are important. Tourists visit India's cities and ride in automobiles, but missionaries who want to reach the village people go to places with no roads. Bullock carts, with large, steel-rimmed wheels like those on the covered wagons of American pioneers, carry people over rough ground, but they are slower than walking. So missionaries walk.

I viewed it as one of my most important jobs to get missionaries back on their feet after accidents, and when Mrs S. arrived at the hospital in Vellore I examined her with great concern. Heat and anxiety had soaked her dress with perspiration, and the awkward angle of her right foot indicated a severe leg fracture. She told me of an accident some months before in which she had broken her thigh bone, the femur. A doctor in the mountains had set the bone, but so far his X-rays had shown incomplete healing. He had sent her to our medical college for examination.

This good woman, Mrs S., politely insisted she had to get back to mission work in her rural area. Legs meant everything to her.

When I X-rayed Mrs S.'s fracture site, I expected to see the wondrously familiar sight of healing bone. Although bone has come to symbolize death at Hallowe'en and in museums, the surgeon knows the symbol lies, for the skeleton is a growing organ. When I cut bone, it bleeds. Most amazing of all, when it breaks, it heals itself. Perhaps an engineer will someday develop a substance as strong and light and efficient as bone, but what engineer could devise a substance that, like bone, can grow continuously, lubricate itself, require no shutdown time, and repair itself when damage occurs?

When bone breaks, an elaborate process begins. Excited repair cells invade in swarm. Within two weeks a cartilage-like sheath called callus surrounds the region and cement-laying cells enter the jellied mass. These cells are the osteoblasts, the pothole-fillers of the bone. Gradually they break down the callus and replace it with fresh bone. In two or three months the fracture site is marked by a mass of new bone that bulges over both sides of the broken ends like a spliced garden hose. Later, surplus material is scavenged so the final result nearly matches the original bone.

That is bone's normal healing cycle. But to my surprise I saw no evidence of this process in Mrs S.'s X-rays. A clean line – a dreaded gap – appeared between the two broken ends of bones, with no mending material fusing them together.

I opened up her leg for a firsthand look and confirmed there was no vestige of healing. Resorting to the inferior, non-living tools of science, I fixed the area of the spiral break with a steel bone plate screwed into both pieces of the bone, above and below. On the other side of the break I placed a grafted section of her tibia to promote new bone formation, then sutured the wound.

After months of casts and wheelchairs and crutches, Mrs

S. again underwent X-rays. They revealed the grafted bone was taking: a milky cloud of growing bone enveloped the new bone strip, joining it to her original femur. But between the ends of the two broken bones a clean division still yawned open. Then I knew we had something very strange. After researching Mrs S.'s history, I learned that twenty years earlier a doctor had irradiated her mid-thigh to treat a small, soft-tissue tumour. Evidently the radiation had killed the tumour. It had also killed all her living bone cells at that site, and thus the two ends would never grow together.

The inactivity was driving Mrs S. crazy. God had sent her to a place where she needed legs, she insisted, and she was determined healing must take place.

I saw one hopeful sign, however: the bone grafts had grown normally. So I performed another operation. I found the space between the two ends of her broken femur so distinct I could insert the edge of my knife and wiggle it. First I checked the steel plate. The two screws farthest from the fracture site were loose and easy to remove: her body had begun rejecting them. But the four screws nearest the fracture were as solid as if they had been drilled into mahogany because the bone there was dead. I had to strain to turn them.

Obtaining two more bone grafts, one from Mrs S.'s other tibia and one from her pelvis, I surrounded her fractured bone with living bone, as if packing it in ice. Then I closed the wound and waited.

Mrs S. recovered and rejoined her mission station in the mountains. She spent an active life trudging the dusty trails, and her improvised leg bone served her well. Seven years later when I had her in for a checkup, X-rays revealed that the original fracture site had never healed – in one little area between the grafted bones I could see light. But a living bone shell, like a huge knot on a tree, had joined the two pieces together and formed a misshapen bulge of bone. She walked entirely on grafts – her original bone above, grafted bone in the middle, and her original bone below.

*

Mrs S. offered a rare example of dead and living bone tissue existing side by side. When I opened her leg, the two looked the same. Their crucial difference showed when living bone interacted organically with her body while the dead bone did not. Because Mrs S. was a living person encountering stresses and forces that required bone renewal, the dead bone failed her. A living body cannot rely on dead bone.

The analogy from physical bone to a spiritual skeleton has already been drawn for us in a dramatic passage in Ezekiel 37. There we see the prophet touring a surrealistic valley piled high with "bones that were very dry" (v. 2). God addressed those bones: "I will attach tendons to you and make flesh come upon you and cover you with skin; I will put breath in you, and you will come to life. Then you will know that I am the Lord" (v. 6).

The bones Ezekiel saw symbolized a great nation, Israel, that had degenerated to the dead skeletal form of antiquity. Israel's faith in God and obedience to Him existed only as a dry, lifeless memory. Yet even those ancient bones had value. Ezekiel watched breathlessly as bones rattled together and formed the framework for a new body. This new nation would come to life with a pre-existing heritage and understanding of God.

The history of a long, personal relationship with God can be preserved in laws and scriptures and ceremonies, as it was with Israel, or in creeds and art and cathedrals, as it is in Western culture today. Some revere such skeletons for their antiquity, buying Mozart's masses and purchasing religious art. But, clearly, the real value of a skeleton only comes to light when it supports a growing organism. Although our laws, scriptures, traditions, and creeds reveal truth in themselves, they exist to serve such an organism, the Body of Christ.

*

The grafted bone in Mrs S.'s leg beautifully displayed the normal procedures of living bone. As surely as the skeleton ossifies and hardens, it simultaneously grows and renews itself. Bone is alive. It spends its days changing, flowing correcting, shifting – like a river as well as a rock.

The same stages of growth that I watched in Mrs S.'s bone graft work faithfully each day within the skeletons of children. The newborn baby has 350 bones which will gradually fuse together into the 206 carried by most adult humans. But many of the baby's bones are soft and pliable, hardly showing the qualities of bone. The birth event would be impossible if a baby were not so compressible and flexible.

As I watch bone ossifying, or becoming hard, in X-rays, I am reminded of my own skeleton of faith. As a newborn Christian my faith was soft and pliable, consisting of vaguely understood beliefs about God and my need for Him. Over time God has used the Bible and other Christians to help ossify the framework of my faith. In the same way that osteoblasts lay down firm new minerals in a bone, the substance of my faith has become harder and more dependable. The Lord has become my Lord; doctrines that were cold and formal have become an integral part of me.

The evangelical wing of faith, especially, tends to convey that all answers can be codified in a comprehensive statement of faith. Those who doubt the basic doctrines are sometimes treated as aliens in the Body and made to grovel in guilt and rejection. For that reason, in the evangelical world doubt is often a private phenomenon. Those of us tempted toward that kind of rigidity must come back to the analogy of living bone. New believers need time for the bones of their faith to strengthen.

I have known many times of doubt. In India I was challenged by the attractions of other religions devoutly practised by millions of people. In medical school I faced constant exposure to assumptions that the universe is based on randomness, without room for an intelligent Designer. As I have grappled with these and other issues – questions about the person of Christ, trust in the Bible, etc. – I have learned it is sometimes helpful to continue accepting as a rule of life something about which I have basic intellectual uncertainties. In other words, I have learned to trust the basic skeleton and use it even when I cannot figure out how various bones fit together and why some are shaped the way they are.

In medical school I was taught by such secular biologists as J. B. S. Haldane and H. H. Woolard, pioneers of evolutionary theory. I noticed that some churches nourished a kind of intellectual dishonesty on this subject. In the university their students took exams and recited the theories of evolution; when they joined the church, they declared their faith in a way that contradicted their exam answers. Ultimately this dichotomy led to a sense of intellectual schizophrenia.

Only after much research and long periods of reflection was I able to put together what I had learned at church and what I had learned at school. But in the meantime I determined that my faith was based on realities that could stand by themselves and that did not need to be subordinated to any explanation of science. Either I would discover that evolution was compatible with the God of my faith, or I would find that evolution was somehow wrong and I would stay with my faith. I operated on that assumption for years during which I was unable to fill in all the blanks about how creation and evolution fit together. (In recent years, new understanding of the nature of DNA has made the possibility of chance evolution so unlikely that the position of one who believes in supernatural intelligence has been tremendously strengthened.)

*

A certain bridge in South America consists of interlocking vines supporting a precariously swinging platform hundreds of feet above a river. I know the bridge has supported hundreds of people over many years, and as I stand at the edge of the chasm I can see people confidently crossing the bridge. The engineer in me wants to weigh all the factors — measure the stress tolerances of the vines, test the wood for termites, survey all the bridges in the area for one that might be stronger. I could spend a lifetime determining whether the bridge is fully trustworthy. But eventually, if I really want to cross, I must take a step. When I put my weight on that bridge and walk across, even though my heart is pounding and my knees are shaking, I am declaring my position.

In the Christian world I sometimes must live like this, making choices which contain inherent uncertainty. If I wait for all the evidence to be in, for everything to be settled, I'll never move. Often I have had to act on the basis of the bones of the Christian faith before those bones were fully formed in me and before I understood the reason for their existence. Bone is hard, but it is alive. If the bones of faith do not continue to grow, they will soon become dead skeletons.

12

ADAPTING

If I profess with the loudest voice and clearest exposition every portion of the truth of God except precisely that little point which the world and the devil are that moment attacking, I am not confessing Christ, however boldly I may be professing Christ.

MARTIN LUTHER

Bone, concealed as it is, does not display its flow of life to onlookers. I must turn to the microscope to see traces of the activity now occurring there. With enough magnification I can identify two types of active cells in bone.

We have already met one type, the osteoblasts, pothole-filling repair cells that attach themselves to fracture sites and lay down bone crystal. But the blasts do not wait around for accidents. Billions of them labour diligently inside me, replacing overaged bone. When I was young, 100 per cent of all the bone in my body was replaced each year. So the jawbone I had as a four-year-old did not contain a single remnant of my three-year-old jaw-bone. Thanks to the wisdom of bone's DNA the shape stayed the same, only larger.

Now only about 18 per cent of my bone gets replaced every year. Old bone does not surrender territory easily, though. It must be dynamited and vacuumed out, and for this job the body has osteoclasts, the demolition team. Clasts are large, packed with an average of ten to thirty nuclei, as if they need all the instructions they can get for their sensitive task.

If I tried to renovate a brick wall by removing a line of bricks in a horizontal row, the entire wall would quickly collapse. If, on the other hand, I removed a brick over here by my left elbow and replaced it, then replaced a brick by my knee, then one up by my head, I could in time safely reconstruct the entire wall. Similarly, clasts scavenge each bit of bone, one cell at a time. They tunnel through bone as easily as moles through a lawn, opening up holes for the blasts to fill. Blasts rejuvenate, depositing a new supply of healthy fibre.

The reckless clast cell leads a kamikaze life, boring through granite with such verve that it burns out after forty-eight hours and is itself escorted away as waste. To me, this cell is employed most beautifully in the bird family. In a brief, crucial span of time, clasts gently invade the bird's bone to loosen up calcium so the mineral can be used to harden the shell of the egg about to be laid.

The blasts and the clasts race throughout a person's life. Blasts tend to dominate the first half, laying down new bone in the orderly design of growth. But demolition clasts eventually outstrip the tired blasts. And so in old age teeth sockets decrease in size, the chin protrudes, the jaw angles in, and the elderly are left with more severe, pointed faces. That is why a fracture causes trauma for the elderly: their blasts, barely up to the rigours of routine repair, heal bones slowly.

As old bone is renewed, blasts factor into their design necessary adjustments for stress. All bone elements are arranged in perfectly engineered, intersecting lines of stress, like the girders on a steel bridge. If I break a foot and the pain of healing makes me adjust my walk so I take shorter steps, gradually those lines of stress in the heel bone will change and will end up at a new angle to the leg. The blasts will accommodate to meet the new challenges.

If I start weight-lifting, a supporting leg bone like the femur might reasonably be expected to buckle or bend. Instead, it becomes thicker and develops extra struts on the stress side. In fact, stress stimulates bone growth. Rest in a hospital bed for a prolonged time and you may lose up to 50 per cent of the calcium in your bones. The astronauts in outer space, freed from gravity, lost up to 20 per cent of their calcium. Walking, lifting, flexing – any activity sends electrical currents through bone to generate growth.

*

When I consider the spiritual Body of Christ, and especially its skeleton of rules governing human behaviour, I am conscious of a parallel type of renewing, adapting activity. The principles God has laid out, sometimes capsulized as in the Ten Commandments and the Sermon on the Mount, do not change, but their specific application certainly changes as the Body of Christ encounters new stresses. Many of the laws and observances of the Bible were geared to a society and culture alien to our own. A continuing need exists for prophets and teachers to interpret unchanging principles in light of the peculiar conditions of their day.

Consider the following list of direct instructions, all given to Christians in New Testament times and recorded for us in the Bible. Some of them are still followed or at least subscribed to by most Christians. Others are practised by only a few denominations who strive to conform literally to New Testament practices. Nevertheless, I know of no group that obeys all of these instructions.

1. Greet one another with a holy kiss (Romans 16:16).
2. Abstain from food sacrificed to idols (Acts 15:29).
3. Be baptized (Acts 2:38).
4. A woman ought to have a veil on her head (1 Corinthians 11:10).
5. Wash one another's feet (John 13:14).
6. It is disgraceful to a woman to speak in the church (1 Corinthians 14:35).
7. Sing psalms, hymns and spiritual songs (Colossians 3:16).
8. Abstain from eating blood (Acts 15:29).
9. Observe the Lord's Supper (1 Corinthians 11:24).
10. Remember the poor (Galatians 2:10).
11. Anoint the sick with oil (James 5:14).
12. Permit no woman to teach men (1 Timothy 2:12).
13. Preach two by two (Mark 6:7).
14. Eat whatever is put before you without raising questions of conscience (1 Corinthians 10:27).
15. Prohibit women from wearing braided hair, gold, pearls, or expensive clothes (1 Timothy 2:9).
16. Abstain from sexual immorality (Acts 15:29).
17. Do not look for a wife (1 Corinthians 7:27).
18. Refrain from public prayer (Matthew 6:5–6).
19. Speak in tongues privately and prophesy publicly (1 Corinthians 14:5).
20. Lead a quiet life and work with your hands (1 Thessalonians 4:11).
21. Lift up holy hands in prayer (1 Timothy 2:8).
22. Give to those who beg from you (Matthew 5:42).
23. Only enrol (for aid) widows who are over sixty, have been faithful to their husbands, and are well-known for good deeds (1 Timothy 5:9–10).
24. Wives, submit to your husbands (Colossians 3:18).
25. Show no partiality toward the rich (James 2:1–7).

26. Owe no man anything (Romans 13:8).
27. Abstain from the meat of animals killed by strangulation (Acts 15:29).
28. If a man will not work, he shall not eat (2 Thessalonians 3:10).
29. Set aside money for the poor on the first day of every week (1 Corinthians 16:1–2).
30. If you owe taxes, pay taxes (Romans 13:7).[1]

A biblical scholar can research those commands he considers occasional and explain why the Bible writer applied the principle in just that peculiar way to a stress. For example, the apostle Paul gave many instructions on eating meat that had passed through heathen ceremonies in the temples, a problem not common today in Western nations. Also, in those days in a church like the one at Corinth, women were judged by powerful social customs. If a woman spoke out in a public meeting, the group would naturally assume she was a prostitute or pagan priestess; the same inference was drawn about women who wore their hair in certain styles.

Paul realized the need to adapt lines of stress depending on what group he was with. He refused to let Jewish Christians force Gentiles to be circumcised unwillingly, yet he went through purification rites in the Jerusalem temple (Acts 21) to win the trust of Jewish Christians.

Today we are facing our own particular stress lines. When the human race was young on a planet of unbelievable expanse and few people, the law "Be fruitful and multiply" was obviously appropriate. But we have obeyed that one command so well that all life is now endangered. We need to place new emphasis on our responsibility for the soil and wildlife and perhaps slow down on our multiplying.

Now that we can separate the enjoyment of sex from the

risk of increasing the number of children, we need new ways to emphasize the Christian view that sex is a means to an end and not an end in itself. If it is not always a step toward the making of a child, how can it be reaffirmed as a symbol of the continuing love that binds a marriage together, and not as an haphazard expression of lust?

Some in the church are trying to adapt to stresses created by the medical profession. When major diseases assailed health, rules to prolong life were developed. Today science has an ability to prolong life almost indefinitely, even when that life is meaningless, without consciousness or hope of recovery, yet not meeting any of the old criteria of death.

These issues do not call for sweeping revisions of creeds and beliefs, but they do evince a need for some members of the church to reflect, study the Bible, and pray, and then lead the way in reinterpreting the will of God for their own generation. These people, prophets and teachers, serve as living bone cells in Christ's body, laying down the inorganic minerals that go into our frame. They should possess humility and a commitment to preserve the great principles of the Christian faith. Yet they should have equal concern that the principles be relevant and give strength just where it is needed.

*

In 1892 Julius Wolff first noticed lines of stress in the cellular arrangement of the human skeleton, leading to Wolff's Law, which every medical student learns. Caught up in his enthusiasm, Wolff declared that bones were in a state of great flux, adapting readily to changes in environment and function. Actually, when I visit a museum and compare skeletons throughout the centuries, I am chiefly impressed by their uniformity. Adaptations to stress are minor knobs and slight ridges along bones that have consistently maintained a definite length and shape.

Behind each of the adaptations of divine law applied to a specific culture in the Bible stands a basic principle. Respect for life must be cherished, although today we redefine life in light of new medical advances. Modesty must be protected, but today a woman's short hair is not immodest. The bone endures; the Body simply adapts to new stresses.

13

INSIDE-OUT

Thy bone is marrowless, thy blood is cold.

MACBETH

Twice a year a strange fever rolls like a fog from the bayous, spreading across the flatlands of Louisiana. Hand-painted signs are propped up outside dilapidated restaurants: FRESH CRAWFISH NOW! Schoolboys, bare-foot and sweaty, scramble up the gullies dragging tin pails crawling with dozens of the prehistoric-looking creatures. Each pail contains a writhing mass of crushed antennaé, flexing pinchers and clicking skeletons.

You can find crayfish (or crawfish) in almost any river, pond, or ditch in Louisiana and in most other states. Highbanked ditches running east and west are the most likely places, since crayfish shy away from the hot sun, and that alignment provides them with more shade. Early morning or evening, squat down beside the creek and wait. Soon your eyes will adjust to the shimmering surface and you can focus on the underwater world. Probably you won't see crayfish right away. They are subtle, and a green or brownish coloration camouflages them masterfully.

As you stare, you gradually see a monster. First are two armoured claws, hinged, hooked at the elbow and menacing-looking. Crayfish claws, half its body length, give it an unbalanced, militaristic appearance, like a gunboat with two oversized howitzers protruding over its bow. Two gleaming black eyes jut out between the claws, eyes that protrude on the ends of stalks – *movable* stalks. If the crayfish wants to see you from a better angle, he does not tilt his head, but jerks his eyestalks as easily as you raise an eyebrow.

If catfish are the garbage collectors of ponds, crayfish are the refuse compactors. Anything goes into their mouths: snails, other crayfish, plants, frogs, fish – living or dead, fresh or carrion. Their elder brothers, the lobsters, contentedly munch hardshell crabs, clams, and mussels. This devouring of things stony is made possible by some ingenious equipment, including two segmented, short limbs called *foot-jaws* which crush and tear whatever is placed between them. Inside, the crayfish stomach features three hard, bony teeth that continue the mastication process.

The rest of the crayfish duplicates in miniature the familiar lobster: plates of overlapping armour ending in a broad, fan-shaped tail.

In 1879 Thomas Henry Huxley wrote a classic book about crayfish. He told of the vile habits of a species that may attack its own babies or devour its spouse after a vigorous mating session. He reported the amazing process of regeneration whereby a crayfish with a missing claw will miraculously sprout a new one. He described the unique qualities of crayfish blood, colourless, which adjusts to the temperature of the surrounding water. The clear liquid draining from a wounded crayfish hardly brings to mind the river of life, but to a crayfish red blood probably seems extravagant.

I write of crayfish not because of their blood or bad tempers or capacity for regeneration, but because of their skeletons. Crack open a crayfish and you'll find soft, white meat, begging to be dipped in butter. No bones grow there to annoy a gourmand – the shell *is* its skeleton. When crayfish season arrives in Louisiana, local restaurants will bring you platters of thirty or so of the boiled creatures, their shells tinted a bright red by the cooking process. After an hour of popping and scraping and digging, you leave a plateful of skeletons – thin, crayfish-shaped exteriors which, if propped up in a lifelike pose, would look like a complete crayfish.

A crayfish has an exoskeleton. Its muscles work against a skeleton surrounding it, and the hardness of the crayfish becomes its chief offence and defence in a competitive world.

*

After devoting several chapters to the essential property of hardness in the bones (doctrines and principles) of Christ's Body, I must, for the sake of balance, insert a strong warning. I sense the need for that warning best when I compare the crayfish and lobster family to human beings. The difference is obvious, especially when you try to shake hands with both. A human feels soft, warm, responsive. If you shake hands with a crayfish, you'll feel inflexibility, coldness, and probably pain. A good-sized lobster can break your finger with a quick pinch of its claw

As I look at the history of the church, failures loom large – failures which can be traced to a misunderstanding of the place of the skeleton in the Body of Christ. Some Christians who realize the importance of law and discipline unfortunately wear their skeletons on the outside. When you meet these people, their dogma stands out as obtrusively as does a crayfish's shell.

Examples leap to mind: the "athletes for God" monks who wished to display their dedication to God publicly and convincingly, Simon Stylites, who died in A.D. 459, led the way: he perched on a pillar east of Antioch for thirty-six years and is said to have touched his feet with his forehead more than 1244 times in succession. Other monks subsisted by eating only grass. Theodore of Sykeon, a seventh-century saint, spent most of his life suspended from a rock in a narrow cage, exposed to the storms of winter, starving himself while soulfully singing psalms.

Some of these practitioners sought a personal way of demonstrating their commitment to God. Others, however, strained to make a public display of their zeal in order to impress onlookers – exactly the error which Jesus blasted in the Pharisees (see Matthew 23 and Luke 11).

Today the most rigorous expressions of faith are seen in the religions of the East, where zealots walk on hot coals and lie on beds of nails. But subtle means of displaying exoskeletons persist in Christianity.

Find a non-Christian walking the streets of your city or town. Pull him aside and ask him what impressions he has of truly committed Christians – not the church-on-Sunday kind but the earnest, born-again kind. Fleeting images will likely cross his mind. He may mention cartoons of the doomsday, sandwich-board prophets who have become a cliché in magazines like the *New Yorker*. He may refer to radio preachers who assail him with threats of hell. Or, he may identify Christians around him by a certain life-style, a list of things they do not do: smoking, drinking, swearing, attending movies, or dancing.

How is an evangelical identified in today's world? Often they are perceived as people who obey strict rules. Psychiatrists excoriate them as guilt-inducers, declaring that over half their patients got messed up in church. Somehow we keep producing variations on pole-sitting Christians. We tend to retreat into our exoskeletons and define our place in the world by how different we are from the rest.

I am often tempted to view legalism as a harmless diversion of the faith. So what if one denomination chooses to ban an innocent activity? Isn't it merely humorous that churches overseas, whose members readily drink and smoke, recoil in horror at the idea of Christians wearing blue jeans or chewing gum? Perhaps some of our cultural quirks are harmless diversions.

But legalism contains enough inherent dangers to elicit the strongest warnings in the Bible. No other issue – not pornography, adultery, violence or the things which most rankle Christians today – inspired more fiery outbursts from Jesus.

Strangely, the people who made Jesus livid with anger were the ones the modern press might call Bible-belt fundamentalists. This group, the Pharisees, devoted their lives to following God. They gave away exact tithes, obeyed each minute law in the Old Testament, and sent out missionaries to gain new converts. Almost no sexual sin or violent crime was visible among the Pharisees. Yet Jesus denounced these model citizens. Why?

To answer that question, I go back to the humble crayfish creeping along the creek bottoms of Louisiana. In comparing its exoskeleton with my more advanced internal skeleton, several differences suggest themselves and throw light on Jesus' strong statements in Matthew 23 and Luke 11 about the dangers of legalism.

First, the crayfish relies almost exclusively on its skeleton for protection. Its dependable armour plating can ward off enemies. Humans, in contrast, have soft, vulnerable exteriors. But as the rules God gave to free His Body begin to calcify, we tend to crouch down inside them for protection. We develop a defensive exoskeleton. In his *Letters to an American Lady*, C. S. Lewis said, "Nothing gives a more spuriously good conscience than keeping rules, even if there has been a total absence of real charity and faith."

Legalists fool you. Like the Pharisees and the "athletes for God," they impress you with their unquestioned dedication. Surely, you think, they have a high view of God. But I learned as I grew up in a legalistic environment that legalism actually errs by lowering sights. It spells out exactly what a person has to do to meet God's approval. In so doing, legalists can miss the whole point that the gospel is a gift freely given by God to people who don't deserve it.

A meticulous researcher named Merton Strommen recently surveyed seven thousand Protestant youths from many denominations, asking whether they agreed with the following statements:

"The way to be accepted by God is to try sincerely to live a good life." More than 60 per cent agreed.

"God is satisfied if a person lives the best life he can." Almost 70 per cent agreed.

"The main emphasis of the gospel is on God's rules for right living." More than half agreed! One would think the apostle Paul and Martin Luther had never opened their mouths, or that Jesus had never come to die. Christians – a majority of young ones – still believe that following a code of rules gets you accepted by God.[1]

What else but our relentless, harping insistence on strict rules could cause this phenomenon? Shouldn't we devote equal times to explaining that rules are merely joints and bones to make our Body effective, and not a ladder to God?

*

A second danger of legalism is that it limits our growth by forming a hard, crusty shell around the accepted group.

An adult crayfish only has the opportunity to grow about once a year. Growth entails an arduous, tortuous procedure called moulting that exposes the creature to deadly dangers. The confining exoskeleton must be shed. Warming up for this traumatic experience, the crayfish rubs its limbs against one another, moves each separately, then flips onto its back, flexing its tail up and down. These movements give it a little play inside its shell.

After several spasms of agitation, the crayfish pushes mightily and its top plate of armour pops free, staying connected only at the mouth. Gingerly, it removes its head, with special care for the eyes and antennae, which are sometimes damaged in the process. Next, legs are yanked out, often with one breaking off. Finally, with a sudden spring forward, the crayfish unsheathes its abdomen and lies there naked and weak.

After resting, prostrate, from the demands of shedding its shell, the crayfish slinks toward some protection. Its body is no longer a stiff, lacquered sheet of chitin; now it has the consistency of wet paper. Often the moulting crayfish will make the discarded skeleton its first meal, ingesting minerals it will need to grow a new shell.

During the next few weeks, the crayfish does all its growing for a year. It may add as much as an inch to its length before the new shell hardens and locks it into the shape and size of the new skeleton.

I have undergone a parallel process of Christian moulting. I started in a close-knit group holding rigid ideas of what a Christian was and who was worthy of fellowship. As I travelled and gained breadth of experience, I realized that not all Christians were of my race with my style of worship and my footnoted doctrinal statement. So I grew a new shell, until the next experience came along. I tended to lapse into seeing the Christian family as an exclusive set of *people like me* encased in a shell. Inside, all was warm and comfortable; outside, the shell protected us from "the world".

But Jesus never described anything resembling an exoskeleton which would define all Christians. He kept pointing to higher, more lofty demands, using words like love and joy and fullness of life – internal words. When someone came to Him for a specific interpretation of an Old Testament rule, usually He pointed instead to the principle behind it.

Jesus understood that rules and governing behaviour are meant to free movement and promote growth as a vertebrate skeleton does, not to inhibit growth as an exoskeleton does.

*

Perhaps the most pernicious effect of legalism is its influence on groups outside the legalistic community. Lobsters and crayfish make unappealing pets because of their external shells. If doctrines and rules are worn externally, as a show of pride in spiritual superiority, the exoskeleton obscures God's grace and love, making the Christian gospel ugly and unattractive.

Earlier in this century in India and other Asian countries, the missionaries' tendency to westernize the church created a hard exoskeleton that offended the local society and limited the church's influence.

In America, too, examples persist. Find a person once deeply involved in church who has chosen to leave it, and you will likely hear that something harsh obtruded into that person's faith. Perhaps it was some Christians' judgmental attitude about a marriage situation. How many divorced persons have left the church when made to feel like second-class citizens? Or perhaps it was disapproval of a habit, like smoking. Having treated emphysema and removed cancerous lungs, I hate smoking. And I hate what divorce does to its victims, especially the children. But I must not allow my views on smoking or divorce to drive people away. For a model, I must look to Jesus, who hated the sin but loved the sinner. Though He openly declared God's laws, somehow He conveyed them with such love that He became known as the friend of sinners.

Do we drive people away from the riches of God's love because of our ideas of what behaviour should be? Rules about behaviour certainly have a function; the Bible swells with them. But they are meant to be worn on the inside, not on the outside as a display of superiority.

A troubling phenomenon recurs among young Christians reared in solid homes and sound churches. After living their early years as outstanding examples of Christian faith, many become spiritual dropouts. Did they fail because they concentrated on the exterior, visible Christian life? Did they learn to mimic certain behaviours, nuances of words, and emotional responses? Crayfish-like, did they develop a hard exterior that resembled everyone else's and conclude such was the kingdom of God, while inside they were weak and vulnerable?

When Christianity is an external exercise, it can be cast aside in the manner of a crayfish flinging off its shell. In fact, many crayfish perish from the moulting ordeal, either because of exhaustion or because of their vulnerability to outside enemies.

An outside shell can seem attractive, trustworthy, and protective. It certainly has advantages over a dead, useless skeleton or over no skeleton at all. But God desires for us a more advanced skeleton that serves as it stays hidden.

SKIN

14

VISIBILITY

What is it, then, this seamless body stocking, some two yards square, this our casing, our facade, that flushes, pales, perspires, glistens, glows, furrows, tingles, crawls, itches, pleasures and pains us all our days, at once keeper of the organs within, and sensitive probe, adventurer into the world outside?

RICHARD SELZER

In India, while leprosy research consumed my time, my wife Margaret trained as an ophthalmologist and became an expert eye surgeon. Because many of the neediest people could not travel to the hospital, she and a team of helpers took a well-stocked mobile unit on monthly circuits into rural areas. On a certain date, a designated building, perhaps a school or an old rice mill, would receive a stream of Indians afflicted with runny eyes or blindness. The staff worked under crude conditions, sometimes in stifling heat, devising an assembly line of treatment. If no building was available, they would even set up portable operating tables under a banyan tree. Sometimes two doctors performed over one hundred cataract operations a day.

In 1956, Margaret's team staffed a camp for several weeks in an area of India that had been devastated by drought. Crops had failed for five years, and the wells were dry of drinking water. People straggled in from every direction, begging for food. Assuming they would have to stay at the camp to receive food, many volunteered for needless surgery – to the extent of asking that one of their eyes be removed – in order to get something to eat.

Young boys volunteered to assist at that hectic camp, and Margaret was assigned a shy, dark boy about twelve years old. He stood on a box, with an impressive but baggy hospital gown wrapped around him, charged with strict instructions to hold a three-battery flashlight so that the light beamed directly on the cornea of the patient's eye. Margaret was dubious: could a young village boy who had never watched any surgery endure the trauma of seeing people's eyes sliced open and stitched together again?

The child, however, performed his task with remarkable aplomb. During the first five operations he scrupulously followed Margaret's instructions on when to shift the angle of light, aiming the beam with a steady, confident hand. But during the sixth case he faltered. Margaret kept saying softly, "Little brother, show the light properly", which he would momentarily do, but soon it again would dangerously bob away from where she was cutting. Margaret could see that he simply could not bear to look at the eye being worked on. She stopped and asked if he was feeling well.

Tears ran down his cheeks and he stuttered, "Oh, doctor – I-I cannot look. This one, she is my mother."

Ten days later the boy's suffering was over. His mother's stitches were removed, and the team gave her eyeglasses. She first tried to blink away the dazzling light, but finally adjusted, focused, and for the first time in her life saw her son. A smile creased her face as she reached out to touch him. "My son," she said, "I thought I knew you, but today I see you." And she pulled him close to her.

*

In her poignant way, the Indian woman expressed that her

son at last had become a recognizable image to her. Before, she had known the sensation of touching him and the sound of his voice. Now she had a literal image of his shape and appearance. If he entered her dreams at night, she would know him. Yet she had seen only one organ of his body, the skin. Our impressions and memories of each other come packaged in that one visible organ by which we judge others and convey our own responses.

Sometimes I envy my wife's field of medicine which is confined to two transparent ovals not shielded by the opaque fabric of skin. She can peer inside without cutting, and if she must cut, can later observe healing with an unobstructed view. Only the eyes expose to the doctor moist, living cells inside the body: corpuscles shooting through capillaries and traces of bacteria and cancer.

Yet in more subtle ways the skin is, like the eye, a window. On it we read the health of the activities within. Anaemia shows in the nails and skin, drawing a ghostly pallor across its victims. Jaundice yellows the skin, while a form of diabetes shades it bronze. Some drugs transform the skin into an iridescent tattoo-blue; we have such patients at Carville. Lack of oxygen in the blood casts a purple tint. Scurvy, beriberi, glandular malfunctions – the skin reveals their presence and many other deficiencies.

When its rainbow is depleted, the skin turns to other signals. Leprosy shows itself when nerve endings fall silent. Cancers leak out in a rash or an aggravating mole. An allergist can crack the secret code of your body's likes and dislikes merely by mapping out a grid on your back and pricking the skin with pin-sized potions. Is it dog hair? Pollen? Shellfish? Your skin will unriddle the mysterious vomiting or sneezing.

*

Skin also provides a window to the emotional world within. We have relatively few voluntary muscles on our skin – we cannot twitch it at will, as can a horse. But we do have control over our faces, and there volumes are written. Pains from childhood are sometimes stamped on to the contours of skin like carved initials scarring a tree trunk. A slight downward curve of the lips can warn a spouse to walk on eggshells.

Sometimes the body revolts and shows its true sentiments in spite of us. Mark Twain said, ''Man is the only animal who blushes – or needs to.'' Blush. It connotes a sudden heat, a steamy swelling of vessels which involuntarily, rebelliously even, rush fifty times more blood to the skin. (Imagine a city water supply responding to an instantaneous 5000-per cent increase in demand.) The young blush more than the old, women more than men. No one is exempt: blind people blush; all races blush, including the darkest ones (their albinos prove it). Blushing acts upon the skin to alert an observer to underlying sentiments.

*

There is no organ like the skin. Averaging a mere nine pounds, it flexes and folds and crinkles around joints, facial crags, gnarled toes, and fleshy buttocks. It is smooth as a baby's stomach here, rough like a crocodile there. A bricklayer's hands may be horny, taut, and layered with sandpaper, but flaccid, pliable folds shroud his abdomen. Intricate spot-welds fasten a leg's wrap, holding it tautly to the muscle layer; an elbow droops loosely, like the skin of a cat that can be tossed by the scruff of its neck.

Choose sections of the scalp, the lip, the nipple, the heel, the abdomen, and the fingertip to view through a laboratory microscope. They are as different as the skins of a host of species – a patchwork somehow growing in a continuous sheet over the body. Tiny ridges crisscross skin's surface to provide traction, much as a snow tyre does. Amazingly, for no apparent reason, each of us is given a different pattern for the ridges, a flourish which the FBI capitalizes on in its fingerprint files. The ridges themselves give texture and the power to grasp a slippery object.

We have a love affair with skin, and our chief response, curiously, is to adorn it. Males perform a daily morning ritual to chop off a night's growth from hair follicles. They rearrange other hairs atop their heads, perhaps worry over a few pimples, and inspect a mole or two. Females expand the ritual, countering the hard work of dense oil glands on the nose by powdering them dry, curling hairs around the eye, plucking others, and outlining that organ with traces of bright colours. Some daub the skin like canvas, hiding it under a paste of colour; most shade the lips to match the day's attire. And then, alone of all the animals on God's earth, we sense the need to swathe large patches of that skin, supporting a multibillion dollar fashion industry in the process.

I can best understand the urge to adorn by studying the competition, the several million other species with whom we share this planet. Choose any class of animal – snakes, insects, birds, mammals – and flip through a colour photograph book of its members. Brightness and design leap from the pages. It is as if the Creator began with exhilaration, splurging on a macaw and a killer whale and a coral snake, took a breather with greyish lizards and dull sparrows and game fish, then lavished tropical fish with just-invented pigments, splashed a cardinal and a jay and a magpie before settling down to the more intricate designs of reptile scales, zebra stripes, and cheetah spots. Then, creativity exhausted and pigment running dry, He settled on blandly uniform flesh colours for the human species –

with interesting variegations of yellows, browns, and reds, to be sure – nevertheless solid and one-coloured except for a wash of coral across lips and nipples.

How skilled is the Creator? Consider the line drawings, paintings, sculptures, and photographs that have, since cave men eloquently expressed our unending fascination with plain human skin.

By studying skin's biochemistry, I can fathom how a few molecules here and there, interacting with sunlight, can change a colour (the Negroid race derives its rich shade from a mere one-thirtieth of an ounce of melanin). I can comprehend the process by which moist, gelatinous cells march to the surface to flatten out and dry into keratin, a protective, flaky coating, before being shed. I can understand the complex process of keratin producing rigid fingernails and horses' hooves. But no amount of training will lessen my astonishment as I watch a single stalk of keratin push its way out of a follicle, grow erect and proud and shockingly unfurl as a peacock feather. What was chemistry becomes beauty. It is as if a brilliant Appalachian quilt springs from a rock, as if a desert suddenly gives birth to a gang of cavorting porpoises.

*

Compared with other finely decorated animals, the human seems naked, vulnerable, incomplete. More than that of any other species, our skin is designed not so much for appearance as for relating, for being touched. And this aspect of skin summons up the basic function of skin within the Body of Christ. In that Body, skin becomes the presence of Christ Himself, the membrane lining that defines our community and enshrouds God's Body in the world. We have seen that Christians sometimes err by displaying their skeletons before a watching world. Christ condemned that trend. Instead, He held before us the principle of love, saying, "All men will know that you are my disciples if you love one another" (John 13:35).

The analogy of skin — soft, warm, and touchable — conveys the message of a God who is eager to relate in love to His creations. Christ was saying to us: Let the world first see the beauty and feel the softness and warmth of the Christian community, and then let it realize the underlying internal framework.

As the world encounters Christ's Body, what is its texture, its appearance and "feel" — its skin? Do people see "love, joy, peace, patience, kindness, goodness, faithfulness, gentleness and self-control"? (Galatians 5:22) We judge people by appearance, studying facial expressions for some hint of mood or glimpse into them. In the same way, we as a Body are being scrutinized and evaluated. Others are drawing a picture of Christ from our appearance. The atmosphere in a church will, skinlike, reveal the substance underneath.

15

PERCEIVING

The greatest sense in our body is the touch sense. We feel,
we love and hate, are touchy and are touched, through the
touch corpuscles of our skin.

J. LIONEL TAYLOR

In 1953 I toured the United States on a Rockefeller
Foundation grant, studying under renowned hand surgeons
and pathologists to explore why leprosy causes paralysis. My
trip ended in New York where I was to speak on assignment
for the American Leprosy Mission and visit several
surgeons. During the meeting with ALM, I began to feel
nauseated and dizzy. I managed to deliver the address, but
the fever continued to rise as I made my way to the subway
station. At one point I swayed and fell to the floor of the
subway car, too dizzy to sit or stand. Other passengers,
probably assuming I was drunk, simply ignored me.

Somehow I staggered to my hotel. I dully realized I
should call a doctor, but the hotel room had no telephone
and the illness overwhelmed me so that all I could do was
curl up on the bed and moan. For several days I stayed
that way, with a bellboy daily fetching me orange juice and
milk and aspirin.

I recovered, though weak and unsteady, in time for my
ship's voyage back to Southampton, England. After landing

at Southampton, I took a train to London, sitting in a cramped corner, hunched over and wishing the interminable trip would end.

At last I arrived at my aunt's house, emotionally and physically drained. I collapsed like a sack of potatoes into a chair and pulled off my shoes. Then came probably the blackest moment of my entire life. As I leaned forward and pulled off my sock, I became aware of the horrible fact that my left heel had no feeling.

A dread fear worse than nausea gripped my stomach. After seven years of working with leprosy patients, had it finally happened? Was I now to be a patient myself?

I stood stiffly, found a straight pin, and sat down again. I lightly pricked a small patch of skin below my ankle. I felt no pain. I jabbed the pin deeper, longing for a reflex, but there was none – just a speck of blood oozing out of the pinhole. I put my face between my hands and shuddered, longing for pain that would not come.

For seven years my team and I had joined in the battle against centuries of tradition to gain new freedom for leprosy patients. We had tried to combat fear, had helped tear down the ugly barbed-wire fence around the leprosy village at Vellore.

I had assured new staff members that the disease was the least contagious of all communicable diseases and that proper hygiene would practically guarantee they would not contract the disease. Now I, their leader . . . a *leper*. That vicious word I had banned from my vocabulary rose up like a monster with new meanings. How glibly I had encouraged patients to overcome the stigma of the past and forge a new life for themselves by surmounting society's prejudices.

My mind clouded. I would have to separate myself from my family, of course – children of patients were the most susceptible group. Perhaps I should stay in England. But what if word somehow leaked out? I could envision the headlines. And what would happen to my leprosy work? How many would now risk becoming social outcasts to help unfortunate victims?

I lay on my bed all night, fully clothed except for shoes and socks, sweating and breathing heavily from tension. Scenes flickered through my mind – poignant reminders of what I would lose as a leprosy patient. Although I knew that sulphone drugs would probably arrest the disease quite quickly, I could not avoid imaging the disease spreading across my face, over my feet, and to my fingers. My hands were my stock in trade. How could I use a scalpel on living organs without exquisite finger control and response to pressure? My career as a surgeon would soon end.

So much of beauty would also slip away. I had always found my greatest relaxation working in a garden. I loved to pulverize the soil with a hoe, then bend down and squeeze it. Crushing the dirt through my fingers brought a wealth of sensations: hardness in the clods, dew on the grass, and a sense of the soil's dampness or claylike qualities. I might lose that sensitivity.

I would no longer feel the pleasing softness of petting a dog, or the flutter of a June beetle cupped in my hands, or the prenatal stirrings of a caterpillar throbbing ominously against a rough cocoon. Feathers, frogs, flowers, wool – touch sensations filled my world. Because I worked with leprosy patients who had lost most of these sensations, I cherished them more consciously than most people.

Dawn finally came, and I arose, unrested and full of despair. I stared in a mirror for a moment, summoning up courage, then picked up the pin again to map out the affected area. I took a deep breath, jabbed in the point – and yelled aloud. Never has a feeling been so delicious as that live, electric jolt of pain synapsing through my body. I fell on my knees in gratitude to God.

I laughed aloud and shook my head at my foolishness of the night before. Of course, it all made perfect sense now. As I had sat on the train, weakened enough to forgo the usual restless motion of muscles in a cramped place, I had numbed a nerve in my leg. Exhausted, I had exaggerated my fears and jumped to false conclusions. There was no leprosy, only a tired, nervous traveller.

*

That dismal experience, which I was too ashamed to mention to anyone for years, imprinted in me profound lessons about pain and sensitivity. Since then I have purposefully tried to feel, *really* feel the extravagant number of objects surrounding me. Forests, animals, cloth, sculpture, paintings – these beg for eager explorations by sense-hungry fingertips.

The skin does not exist merely to give the body an appearance. It is also a vital, humming source of ceaseless information about our environment. Most of our sense organs – the ears, the eyes, the nose – are confined to one spot. The skin is rolled thin like pie dough and studded with half a million tiny transmitters, like telephones jammed together waiting to inform the brain of important news.

Think of the variety of stimuli your skin monitors each day: wind, particles, parasites, changes in pressure, temperature, humidity, light, radiation. Skin is tough enough to withstand the rigorous pounding of jogging on asphalt, yet sensitive enough to have bare toes tickled by a light breeze. The word *touch* swells with such a plethora of meanings and images that in many dictionaries, including the *Oxford English*, its definition runs the longest of any entry. I can hardly think of a human activity – sports, music, art, cooking, mechanics, sex – that does not vitally rely on touch. (Perhaps pure mathematics?)

Touch is the most alert of our senses when we sleep, and it is the one that seems to invigorate us emotionally: consider the lovers' embrace, the contented sigh after a massage, the cuddling of a baby, the sting of a hot shower. Read the thoughts of Helen Keller – a *cum laude* graduate of Radcliffe and author of twelve books – and you will see what the brain can accomplish with no input but a sense of touch.

Although scientists disagree on exactly how touch works, they can calibrate how well it works. One tap of the fingernail can tell me if I am touching paper, fabric, wood, plastic, or steel. A normal hand can distinguish between a smooth plane of glass and one etched with lines only 1/2500 of an inch deep. A textile feeler can readily recognize burlap by the friction – that's easy; but he can also pick satin over silk, blindfolded. By rubbing his hands over a synthetic fabric, he can detect if the nylon blend has been increased by 5 per cent.

Those seemingly useless hairs blanketing our bodies act as levers to magnify the sensation of touch. We can discern a thousandth of an ounce of pressure on the tip of a half-inch hair.*

*

The skin's advanced ability to inform helps me understand one of the chief duties of the front line of the Body of Christ:

---

* A good friend of mine, Dr Khonalker of Bombay, learned firsthand about the sensitivity of hairs when he tested a group of women to seek the normal threshold of sensitivity. He discovered that the women who did not usually shave their legs were very insensitive in the area shaven for testing for a while, as insensitive as leprosy patients. But gradually their skin adapted. The skin had a potential for greater sensitivity, but suppressed it as long as hairs were present. When shaving removed the hairs, the body noticed silent areas on the legs and "turned up the volume" on touch-sensitive cells there.

to sensitively perceive the people it contacts. Inexperienced counsellors, eager to help people, are warned, "First, you must listen. Your wise advice will do not good unless you begin by carefully listening to the person in need." Skin provides a more basic kind of listening, a tactile perception from thousands of sensors. Love for others starts with this primal contact.

If there is a change in the air pressure, or in the texture of cloth, or in the temperature, skin sensors fire off messages to the brain. In the same way the Christian church being, as Jesus said, "in the world but not of the world", encounters a constant stream of signals about the qualities and needs of its environment. The Body is large, universal, and its sensors report in simultaneously from Marina Towers in Chicago, the slums of Harlem, the jungles of Peru and Sri Lanka, and the deserts of Russia and Arabia.

In Christ's Body, some members are specifically designated to monitor the changing needs of the world. Today, for example, Christian missions are becoming more sensitive to the physical and social needs of people, as well as their spiritual needs.

The early days of missions were sometimes marked by people who were unresponsive to new environments. They did not sense the worth and beauty already present in strange cultures. They responded to bare-breasted, drum-beating Africans as if they were inchoate Europeans, swaddling them in inappropriate clothes and teaching them Martin Luther's favourite hymns.

Condescending love intrudes on the scene with a slick solution devised in a place far away from the human need. The best, most effective kind of love begins with a quiet listening, a tactile awareness.

*

Of all the senses, touch is most trustworthy. A baby first relates to the world through the sense of touch. Give him an object to play with and he will finger it, then bring it to his mouth and tongue it. To him, auditory and visual senses are secondary; not until later will he primarily value visual sense. But even we adults somehow believe our tactile senses more readily. "Tangible" proof is easier to accept. Thomas doubted visual reports of Christ's resurrection, declaring, "Unless I see the nail marks in his hands and put my finger where the nails were, and put my hand into his side, I will not believe it" (John 20:25).

A child touches a magician's props to see if they are real – he cannot trust his eyes. A mirage can fool the eye and the brain, but not the skin's touch.

I recall one incident when my daughter Mary, three years old, was trying to overcome a fear of violent thunderstorms. She understood we were safe inside our house, and yet as the lightning streaked closer and closer, she ran to me and put her tiny hand in mine. "We aren't afraid, are we, daddy?" she said in an uncertain voice.

Just then a tremendous clap of thunder crashed nearby and all our lights went out. Mary, breathing in short gasps, fearfully cried out, "Daddy! We aren't afraid, are we?" Her words were brave, but I could feel her true thoughts in her stiffened hand, trembling with fear. Skin communicates to skin.

We shall observe qualities of skin that allow for quick adaptation to changes. But these qualities are useless if the receptors of skin are numbed. God Himself chose to establish a tangible presence in the world where He, like men and women, felt – through skin – fatigue, pain, and ultimately death. No better model of tactile love exists than His Son. And now we are called to be His sensitive "skin" in the world.

16
COMPLIANCY

No one but kings and princes should have the itch, for the sensation of scratching is so delightful.

KING JAMES I

As an intern in London I had the great privilege of training under Dr Gwynne Williams, a surgeon who unfailingly emphasized the human side of medicine. He strolled through the halls of poorly heated hospital wards with his right arm Napoleonically tucked inside his coat which, unknown to his patients, concealed a hot-water bottle.

"You can't rely on what patients tell you about their intestines," Dr Williams would admonish us interns. "Let their intestines talk to you." The hot-water bottle made his hand a better listener. He taught us to kneel by a patient's bedside and gently slip a warm hand under the sheets on to the person's belly. "If you stand," he explained, "you'll tend to feel only with the downward-pointing fingertips. If you kneel, your full hand rests flat against the abdomen. Don't start moving it immediately. Just let it rest there."

We learned to feel an instant tightening of the patient's abdominal muscles – a protective reflex. A cold hand assured those muscles would remain tight, but a warm, comforting hand could coax them to relax. We gently caressed the abdomen, earning tactile trust. Once muscles had softened, we could sense the organs' movement, responding to the simple act of breathing.

Dr Williams was right: there is no need to ask questions. A trained hand gently exploring the abdomen can detect tautness, inflammation, and the shape of tumours that more complicated procedures merely confirm. Touch is my most precious diagnostic tool.

*

We have called touch a "basic" sense, but that word can mislead. Actually touch is one of our most complex senses.

Every square inch of the body has a different response to touch. Scientists such as Maximilian von Frey have mapped the nerves as meticulously as Rand McNally has mapped the world. Von Frey measured the threshold of touch, the amount of gramme weight it takes for a person to sense that an object has come in contact with the skin. The soles of the feet, thickened for a daily regimen of abuse, do not report in until a weight of 250 milligrammes per square mm. is applied. The back of the forearm is triggered by 33 milligrammes of pressure, the back of the hand by 12 milligrammes. The really sensitive areas are the fingertips (3 milligrammes) and the tip of the tongue (2 milligrammes).[1]

All nerves seem sluggish compared to those in the cornea of the eye, transparent, deprived of blood and thus incredibly vulnerable. The cornea fires off a response if just two-tenths of a milligramme of pressure is applied. A stray eyelash can make a baseball pitcher stop the game – he can concentrate on nothing else. In contrast an eyelash on his forearm would go unnoticed. Similarly, a wise mosquito will land on the forearm, not the sensitive hand, to go undetected. And only a foolhardy insect would attempt a secret landing on soft lips.

Touch distribution was not handed down at a blackjack table ("God does not play at dice," said Einstein): the sensitivity of each square inch is programmed to fit the function of that body part. Our fingertips, tongues, and lips are the portion of the body used in activities that need the most sensitivity.

Scientists compose their charts based on "normal" reactions to stimuli. Actually, touch changes constantly with its environment. The skin, for example, responds by adapting. A 100 milligramme weight is lowered onto my forearm. Blindfolded, I realize that something is touching me. The sensation stays for four seconds, then fades. My nervous system has adapted; I no longer notice the weight. My body has filtered out the messages coming from nerve endings on my forearm, deciding there is no evident danger and no need to clog up the circuits with useless information about weights on my arm. Involuntarily, I lose awareness of the weight – that is, until the weight is removed, at which time my brain will clearly report a change on my forearm. Were it not for this extraordinary volume switch through which my sensations pass, I could not wear wool or other coarse clothing – my body would constantly remind me of its scratchy presence, and I could hardly concentrate on anything else.

I experience skin's adaptation whenever I lower myself into a hot bathtub. I run the water so hot I can barely stand it and gradually lower my body, first reacting as if I am easing myself into a patch of stinging nettles. Within ten seconds my body has adjusted, and the same water actually feels soothing and comfortable. I can continue raising the temperature of the water, and my body will adapt – up to a maximum point of 115°F., beyond which I will feel constant, nonadapting pain.

*

What prompts such a complex system of perception? Is the brain starved of sensation, desiring a circuit-jamming report from the world outside? Do the body's senses accumulate information out of curiosity? No, the purpose of all this is to prepare the body parts to respond wisely.

The elaborate mechanisms producing touch, for instance, prepare the skin to adapt to changing surfaces. Bioengineers use the word compliancy to denote this response. Compliancy describes the capacity of skin to flow around whatever surface it contacts, a quality skin exhibits better than any comparable material. Compliancy gives the body freedom to move around in any environment, to expose itself to changing conditions and yet keep an unbroken, protective surface. Leather clothes and shoes are highly valued (and priced) because they, as animals skins, have flexibility and elasticity as well as an ability to "breathe". The DuPont Company spent millions of dollars to develop Corfam as a possible substitute for leather, but finally called off their project in failure. Their inventions did not have the compliancy of even dead skin (leather).

In the last few years, as I have tried to design shoes and tools for the feet and hands of leprosy patients who lack the basic sensitivities of touch, I have spent hundreds of hours researching the anatomy of living skin. Underneath the skin in the palm of the hand lie globules of fat with the look and consistency of tapioca pudding. Fat globules, so soft as to be almost fluid, cannot hold their own shape, and so they are surrounded by interwoven fibrils of collagen, like balloons caught in a supporting rope net. Collagen occurs in greater quantity where it is most needed – in those parts that need structure and support. The cheeks and the buttocks have more fat and less collagen, as anyone who has struggled with a double chin or sagging figure unfortunately knows. But where stress occurs, such as on the palm of the hand, fat is tightly gathered and enveloped by fibrous tissue in a design resembling fine Belgian lace.

I grasp a hammer in the palm of my hand. Each cluster of fat cells changes its shape in response to the pressure. It yields but cannot be pushed aside because of the firm collagen fibres around it. The resulting tissue, constantly shifting and quivering, becomes compliant, fitting its shape and its stress points to the precise shape of the handle of the hammer. Engineers nearly shout when they analyse this amazing property, for they cannot design a material that so perfectly balances elasticity with viscosity.

If my skin tissue had been made harder, I might insensitively crush a goblet of fine crystal as I hold it in my hand; if softer, it would not allow a firm grip. When my hand surrounds an object – a ripe tomato*, a ski pole, a kitten, another hand – the fat and collagen redistribute themselves and assume a shape to comply with the shape of the object being grasped. This response spreads the area of contact, preventing localized spots of high pressure, limiting stress while giving firm support.

Take a clattery hand of bone – such as the hand of a human skeleton on display in a biology classroom – and wrap it around a hammer. Against such a hard surface, the hammer handle will contact only about four pressure points. Without my compliant skin and its supporting tissues, those four pressure points would inflame and

* Tomatoes demonstrate the superior qualities of human skin. Commercial tomatoes are bred with thick skins so that insensitive mechanical pickers, which do not have compliant surfaces, won't destroy them. I exult in my own vine-ripened tomatoes, with thinner skins and better taste, which I can pick when they are most succulent because my skin is so compliant.

ulcerate if I pounded the hammer just a few times. But because of compliancy, my entire skin-covered hand absorbs the impact.

Compliancy, a word with special meaning to my engineering colleagues, is a good word biologically and a pregnant word spiritually. I need the inflexibility of my skeleton to keep me upright and to impose my will on the environment, but when I grasp something, it is good that my bones do not contact the object. The compliant tissues covering my bones assume the shape – awkward or smooth – of the object. I do not demand that the object fit the shape of my hand; my hand adapts, distributing the pressure.

The art of Christian living, I believe, can be glimpsed in this concept of compliancy. In daily activity as my shape moves into contact with other, foreign shapes, how does my skin respond? Whose personality adapts? Do I, as does my grasping hand, become square to those things that are square, round to those things that are round? The apostle Paul finished the analogy for us in 1 Corinthians 9:19–22: ''Though I am free and belong to no man, I make myself a slave to everyone, to win as many as possible. To the Jews I became like a Jew, to win the Jews. To those under the law I became like one under the law (though I myself am not under the law), so as to win those under the law. To those not having the law I became like one not having the law (though I am not free from God's law but am under Christ's law), so as to win those not having the law. To the weak I become weak, to win the weak. I have become all things to all men so that by all possible means I might save some.''

17
TRANSMITTING

*In the midst of bloody persecution under Idi Amin's rule in
Uganda, a missionary society in England wrote to a bishop
there, "What can we send your people?" The answer came
back: Not food, not medicine; 250 clerical collars. This was
the explanation: "It is your Western prejudice which thinks
this an odd request. You must understand, when our people
are being rounded up to be shot, they must be able to spot
their priests.*

PAUL SEABURY

Dr Harry F. Harlow loved to stand by the animal cages
in his University of Wisconsin laboratory and watch the
baby monkeys. Intrigued, he noticed that the monkeys
seemed emotionally attached to cloth pads lying in their
cages. They caressed the cloths, cuddled next to them, and
treated them much as children treat a teddy bear. In fact,
monkeys raised in cages with cloths on the floors grew
huskier and healthier than monkeys in cages with wire-mesh
floors. Was the softness and touchability of the cloth an
important factor?

Harlow constructed an ingenious surrogate mother out
of terry cloth, with a light bulb behind her to radiate heat.
The cloth mother features a rubber nipple attached to a
milk supply from which the babies could feed. They adopted
her with great enthusiasm. Why not? She was always

comfortingly available, and, unlike real mothers, never roughed them up or bit them or pushed them aside.

After proving that babies could be "raised" by inanimate, surrogate mothers, Harlow next sought to measure the importance of the mother's touchable, tactile characteristics. He put eight baby monkeys in a large cage that contained the terry cloth mother plus a new mother, this one made entirely out of wire mesh. Harlow's assistants, controlling the milk flow to each mother, taught four of the babies to nurse from the terry cloth mother and four from the wire mesh mother. Each baby could get milk only from the mother assigned to it.

A startling trend developed almost immediately. All eight babies spent almost all their waking time (sixteen to eighteen hours per day) huddled next to the terry cloth mother. They hugged her, patted her, and perched on her. Monkeys assigned to the wire mesh mother went to her only for feeding, then scooted back to the comfort and protection of the terry cloth mother. When frightened, all eight would seek solace by climbing on to the terry cloth mother.

Harlow concluded, "We were not surprised to discover that contact comfort was an important basic affectional or love variable, but we did not expect it to overshadow so completely the variable of nursing; indeed the disparity is so great as to suggest that the primary function of nursing is that of insuring frequent and intimate body contact of the infant with the mother. Certainly, man cannot live by milk alone."[1]

In other experiments, some baby monkeys were raised in cages with only a wire mesh mother. They, too, approached her only for feeding, and many of these babies did not survive. Those who did reacted to stress by cowering in a corner, screaming, or by hiding their faces under their arms.

*

Anthropologist Ashley Montagu reports on these and many other similar experiments in his elegant and seminal book *Touching*. He found close physical contact with a mother animal to be essential to the normal development of young animals. Except for man, all mammals spend great amounts of time licking their young. Animals will often die if they are not licked after birth; they never learn to eliminate waste, as one consequence. Montague concludes that the licking is not for cleanliness, but for essential tactile stimulation.

As pet owners know, animals do not outgrow the urge to be touched. A cat arches its back and brushes it gently across its owner's leg. A dog wriggles on the carpet, begging for a belly scratching. A monkey meticulously grooms and combs the hair on its fellow tribe members.

Montagu even suggests that human foetuses need the massive tactile stimulations of labour. Only the human species goes through such a long, arduous birth process. Montagu believes the fourteen hours or so of uterine contractions, which have been described from the mother's viewpoint but never from the foetus's, may be important stimuli to finish off maturation of certain body functions. Could this explain, he wonders why babies delivered by Caesarean section have a higher death rate and a greater incidence of hyaline membrane disease?[2]

Although the role of tactile stimulation during birth remains speculative, the need for touching after birth has been dramatically, and tragically, demonstrated. As late as 1920, the death rate among infants in some foundling hospitals in America approached 100 per cent. Then Dr

Fritz Talbot of Boston brought from Germany an unscientific-sounding concept of "tender loving care". While visiting the Children's Clinic in Düsseldorf, he had noticed an old woman wandering through the hospital, always balancing a sickly baby on her hip. "That," said his guide, "is Old Anna. When we have done everything we can medically for a baby and it still is not doing well, we turn it over to Old Anna, and she cures it."

When Talbot proposed this quaint idea to American institutions, administrators derided the notion that something as archaic as simple touching could improve their care. Statistics soon convinced them. In Bellevue Hospital in New York, after a rule was established that all babies must be picked up, carried around, and "mothered" several times a day, the infant mortality rate dropped from 35 per cent to less than 10 per cent.

Despite these findings, even today touching is often viewed as an unavoidable part of the more important tasks of feeding and cleaning the baby. Seldom is it considered an essential need in itself without which a baby may never mature. Jewish people are highly tactile, as are Latins, but Anglo-Saxons and Germanic people are low on the scale. In general, though, the higher the social strata, the less parents touch their infants. Perhaps we have reached the extreme in America, where mothers carry their babies at arms' length in plastic carriers and fathers spend an average of thirty seconds per day in tactile contact with their children.

Among some severely disturbed children, such as autists, forceful and persistent touching may represent the only hope for a cure. An autistic child needs almost constant touching and rubbing to trigger a release from his self-hugging isolation.

Montagu decisively concludes that the skin ranks highest

among the sense organs, higher even than eyes or ears. Skin not only conveys information about the world, but also perceives basic emotions. Am I loved and accepted? Is the world secure or hostile? The skin osmotically absorbs these concepts and the world view they provide.

Words of touch have edged into our vocabulary as expressions of the way we relate to others. We rub people the wrong way, or conversely, we stroke them. A gullible person is a soft touch; a volatile one we handle with kid gloves. We are thin-skinned, thick-skinned; we get under each other's skins. We relate tactfully or tactlessly.

The intimate moment of the sex act is our most massive cutaneous experience. We touch so fervidly that two organisms become, for an instant, one. And in the West, a visual-oriented culture, some express a need for sex (often mistakenly equated with love) by exposing larger areas of skin, as if the daring wearer is begging to be touched.

As we grow older, skin offers us the most natural medium for communicating basic emotions, such as love. It is our chief organ of contact with others. Skin cells offer a direct path into the deep reservoir of emotion we metaphorically call "the human heart".

Touching includes risk. It can evoke the cold, armourlike resistance of a hurt spouse refusing to be comforted or the lonely shrug of a child who insists, "Leave me alone!" But it can also conduct the electric tingling of love-making, the symbiosis of touching and being touched simultaneously. A kiss, a slap on the cheek – both are forms of touching, and both communicate.

*

The skin of the Body of Christ, too, is an organ of communication: our vehicle for expressing love.

I think back on how Jesus acted while inhabiting a human

body on earth. He reached out His hand and touched the eyes of the blind, the skin of the person with leprosy, and the legs of the cripple. When a woman pressed against Him in a crowd to tap into the healing energy she hoped was there, He felt the drain of that energy, stopping the noisy crowd and asking, "Who touched Me?" His touch transmitted power.

I have sometimes wondered why Jesus so frequently touched the people He healed, many of whom must have been unattractive, obviously diseased, unsanitary, smelly. With His power, He easily could have waved a magic wand. In fact, a wand would have reached more people than a touch. He could have divided the crowd into affinity groups and organized His miracles – paralysed people over there, feverish people here, people with leprosy there – raising His hands to heal each group efficiently, en masse. But He chose not to. Jesus' mission was not chiefly a crusade against disease (if so, why did He leave so many unhealed in the world and tell followers to hush up details of healings?), but rather a ministry to individual people, some of whom happened to have a disease. He wanted those people, one by one, to feel His love and warmth and His full identification with them. Jesus knew He could not readily demonstrate love to a crowd, for love usually involves touching.

In an earlier chapter (chapter 7) I mentioned the need for us as Christ's Body to devote resources to aid the whole Body by distributing food and medicine throughout the world. Having been at the front lines of such activity overseas, I firmly believe such love is best expressed person to person, through touch. The further we remove ourselves from personal connections with people in need, the further we stray from the ministry Jesus modelled for us.

In India, when I would treat a serious case and prescribe

some drug, sometimes the relatives of the patient would go and purchase the medicine, then bring it back and ask me to give it to the patient "with my good hands". They believed the medicine was more able to help the patient if it was given by the hand of the physician.

*

I live on the grounds of the only leprosarium in the continental United States. Carville has a vivid history. The hospital began after the Civil War when an order of Catholic nuns, the Daughters of Charity, expressed a specific calling to serve leprosy patients. Because no one wanted to live near a leprosarium, a remote plot of swampland was purchased on the Mississippi River under the guise of establishing an ostrich farm. Early patients were smuggled in at night on coal barges, blackened and hiding under tarpaulins.

Word about the leprosarium soon leaked, however, and immediately construction workers quit their jobs. Misconceptions of the disease struck such fear that no one would risk exposure to it. But a calling is a calling, the nuns decided. Under the direction of a stout and courageous Mother Superior, they took up the hoes and shovels themselves, digging canals to drain the swamp. With no prior construction experience, teams of sisters in starched, sweltering habits, dug foundations and erected buildings. Only they cared enough to touch and treat the disfigured patients who came to them in the darkness of night.

Now, nearly a century later, I treat leprosy patients at that same hospital. For many of them, whatever they touch – furniture, fabric, grass, asphalt – feels the same. When they put their hands on a hot stove because it feels the same as a cool one, I must treat their damaged hands.

I hate leprosy. Victims who are not treated feel the disease slowly creep over the hands and feet and they then experience the worst effect of all: they lose the ability to sense human contact. Many cannot even feel when another person holds their hands or caresses them. Because of ignorance and superstition, this disease destroys social contact between victims and their friends, employers, and neighbours. Leprosy is a devastatingly lonely disease.

As at Carville, many of the great advances in leprosy research have come about because of Christian action, especially by the Leprosy Mission and its counterpart, the American Leprosy Mission. I have sometimes wondered why leprosy merits its own mission; I know of no "Malaria Mission" or "Cholera Mission". I think the reason is the starving need of leprosy patients for human touch. It is a unique and terrible need, and Christian love and sensitivity meet it best.

Medical teams at places like Vellore, India, can do great things for leprosy patients. They treat the raw ulcers and painstakingly reconstruct feet and hands through tendon transfers and plastic surgery. They transplant new eyebrows to replace missing ones, repair useless eyelids, and sometimes even restore sight. They train patients in constructive jobs and give them new life.

But of all the gifts we can give a leprosy patient, the one he values most is the gift of being handled and touched. We don't shrink from him. We love him with our skin, by touch.

18
LOVING

Christianity is not a statistical view of life.
MALCOLM MUGGERIDGE

A simple woman named Mother Teresa has been awarded
a Nobel Peace Prize for her work in Calcutta among
members of India's lowest caste. She cannot save all India,
so she seeks the least redeemable, the dying. When she finds
them, in the gutters and garbage dumps of Calcutta's alleys,
she brings them to her hospital and surrounds them with
love. Smiling women daub at their sores, clean off layers
of grime, and swaddle them in soft sheets. The beggars,
often too weak to talk, stare wide-eyed at this seemingly
misdirected love offered so late in their lives. Have they
died and gone to heaven? Why this sudden outpouring of
care — why the warm, strengthening broth being gently
spooned to their mouths?

A newsman in New York — properly outfitted in a three-
piece suit, taking cues from an off-camera Teleprompter
— confronted Mother Teresa with a similar line of
questioning. He seemed pleased with his acerbic probing.
Why indeed should she expend her limited resources on
people for whom there was no hope? Why not attend to
people worthy of rehabilitation? What kind of success rate
could her hospital boast of when most of its patients died

in a matter of days or weeks? Mother Teresa stared at him in silence, absorbing the questions, trying to pierce through the facade to discern what kind of a man would ask them. She had no answers that would make sense to him, so she said softly, ''These people have been treated all their lives like dogs. Their greatest disease is a sense that they are unwanted. Don't they have the right to die like angels?''

*

Malcolm Muggeridge, who wrote a book on Mother Teresa, struggled with these questions also. He observed the filth and poverty of Calcutta and returned to England to write about it with fire and indignation. But, he comments, the difference between his approach and Mother Teresa's is that he returned to England . . . while she stayed in Calcutta. Statistically, he admits, she does not accomplish much by rescuing a few stragglers from a cesspool of human need. Then he concludes with the statement, ''But then Christianity is not a statistical view of life.''

Indeed it is not. Not when a shepherd barely shuts the gate on his ninety-nine before rushing out, heartbroken and short of breath, to find the one that's missing. Not when a labourer hired for only one hour receives the same wage as an all-day worker (Matthew 20:1–16). Not when one rascally sinner decides to repent and ninety-nine upstanding citizens are ignored as all heaven erupts in a great party (Luke 14:4–7).

Christian love, *agape*, giving love, is not statistical either. Perception by the skin is more basic than perception through an eye or an ear. It senses a need and responds instinctively, personally.

I do not believe mission work necessarily becomes more effective as it grows more specialized and impersonal. Sometimes the increase in technology may be inevitable, as in a Christian medical college, but I have watched good Christian medical agencies in India gradually lose their original mission as they become institutionalized, with buildings and staff to support. The "quality" of their treatment rises, but so does the expense. To make the work more self-supporting, they branch out into techniques that attract patients who can pay. Meanwhile, the poor and unloved, who can no longer afford the mission hospital, must turn to the government clinic for help.

In contrast, I look at the impact my parents had. Although they went to India to preach the gospel, by living in tactile awareness of people's needs they began to respond on several levels. Within a year they were involved in the fields of medicine, agriculture, education, evangelism, and language translation. Their perception of needs determined the form (compliancy) their love assumed.

My mother and father worked for seven years before anyone converted to Christianity, and, in fact, that first conversion came as a direct result of their healing love. Villagers would often abandon their sick outside our home, and my parents would care for them. Once when a Hindu priest was dying of influenza, he sent his own frail, sickly, nine-month-old daughter to be raised by my parents. None of his swamies would care for the sick child; they would have let her die. But my parents took her in, nursed her to health, and adopted her as their own. I gained a stepsister, Ruth, and my parents gained an unexpected response of trust. The villagers were so moved by this example of Christian love that a few soon accepted Christ's love for themselves.

Years later, when my mother, Granny Brand, was eighty-five, long after my father had died, she helped forge a medical breakthrough. She had often treated gross abscesses on the legs of mountain people by draining the

pus and excising a long, thin guinea worm. Distressed by
the frequency of those abscesses, she studied the problem
and learned that the worm's life cycle included a larval stage
spent in water. If she could help break that cycle, she would
eradicate the worm. Knowing the people's habits well, she
quickly deduced that wading in water was probably the
means of transmission. Cashing in on the trust and love
she had built up through decades of personal ministry, she
rode her horse from village to village, urging the people
to build stone walls around their shallow wells and to
prevent foot contact with the water. In a few years this old
lady had singlehandedly caused the eradication of all such
worms, and their resulting abscesses, in two mountain
ranges.

*

My wife Margaret had a similar experience with a terrible
condition afflicting the eyes of children. Whenever she
encountered this condition, I could read it in the despair
on her face that night. I would look at her and
sympathetically murmur one word, "Keratomalacia?" and
she would nod "yes".

The condition resulted from a deficiency of vitamin A
and protein among young children between one and two
years old. A baby would be well-nourished as long as it
was breast-fed, but soon a new brother or sister would push
it from its mother's breast. A steady diet of rice failed to
provide needed vitamins, making small bodies especially
susceptible to infection. Finally, an attack of conjunctivitis
– usually one of the easiest infections to treat in a well-
nourished person – would begin to eat away at the
malnourished child's eyes. Looking into those eyes, we
would see a jellied mass of softness and sogginess, as if a
strange heat ray had melted all the parts. Contact with one

of those children, fearfully squinting to keep out light, never failed to overwhelm Margaret, regardless of how many successful procedures she had performed that day.

Then spurred by Margaret's sense of need, some medical college researchers discovered that a common green herb, which grew wild all over our area, contained a remarkably high concentration of vitamin A. They also realized that peanuts, a local crop grown for peanut oil, possessed the missing protein. After mashing the nuts to produce oil, the villagers had been feeding the peanut residue to their pigs.

Now the task became one of education. Margaret and public health nurses spread the word, and soon mothers were excitedly telling their neighbours that the green herb and peanuts could prevent their children's blindness. The news travelled like gossip through the villages, soon protecting children from the dreaded keratomalacia.

These two examples are unusual, of course. Much of mission work consists of exhausting labour with less dramatic results. But they both demonstrate possible results of tactile Christian love. Government data banks, advanced hospitals, and agricultural experts had sufficient knowledge to attack keratomalacia and the guinea worm, but they had not gained the trust of villagers. Impetus for a medical advance came, instead, from workers who were "in touch with" the suffering people and who had built up enough trust and respect to effect the remedy.

An old Chinese proverb says: "Nothing can atone for the insult of a gift except the love of the person who gives it." If I go up to a man who looks poor, press a ten-dollar bill in his hand, and walk away, I really am insulting that person. My action says, "You can't take care of yourself – here's a gift for you." But if I involve myself in his life, recognize his need, and stand alongside him, sharing what resources I have with him, he is not offended.

I wonder how effective Granny Brand would have been had she dropped leaflets from an aeroplane explaining the need for stone walls around wells.

*

Every week my mailbox bulges with appeals for help from Christian organizations involved in feeding the hungry, clothing the naked, visiting the prisoners, healing the sick. They describe to me the horrible condition of a hurting world and request my money to help relieve the pain. Often I give, because I have lived and worked among the world's suffering and because I know most of these organizations conscientiously shed love and compassion abroad. But it saddens me that the only thread connecting millions of giving Christians to that world is the distant, frail medium of direct mail. Ink stamped on paper, stories formula-edited to achieve the best results – there is no skin involved, no sense of touch.

If I only express love vicariously through a cheque, I will miss the incredible richness of response that a tactile loving summons up. Not all of us can serve in the Third World where human needs abound. But all of us can visit prisoners, take meals to shut-ins, and minister to unwed mothers or foster children. If we choose to love only in a long-distance way, *we* will be deprived, for skin requires regular contact if it is to remain sensitive and responsive.

Again, the best illustration of this truth is Jesus Christ, the embodiment of God living on this planet. The Book of Hebrews sums up His experience on earth by declaring that we now have a leader who can be *touched* with the feelings of our weaknesses (Hebrews 4:15). God Himself saw the need to come alongside us, not just love us at a distance. How could He fully manifest love except through human flesh? Jesus is said to have "learned obedience from what he suffered'' (Hebrews 5:8). A stupefying concept: God's Son learning through His experiences on earth. Before taking on a body God had no personal experience of physical pain or of the effect of rubbing against needy persons. But God dwelt among us and touched us, and His time spent here allows Him to more fully identify with our pain.

The ideal, then, is to give love to someone you can touch – a neighbour, a relative, a needy person in your community. I was able to do that in India. Now, I look for people in Carville to love through touch. Of course, I still have great concern for the needs of the people of India, and I seek out others who can love the Indians through touch. I support these people and their organizations with my gifts and prayers.

Touch can be secondhand, both in the body and in the church. Touch corpuscles are located deep inside my skin, and the activities on the surface can indeed reverberate through other cells, conveying the sense of touch. I give to India through my daughter in Bombay, through my friend at the leprosy hospital in Karijiri, Dr Fritchie, and through other people like them. They apply my love in person, and I expect from them a sensitive report on the results of that love. It is my responsibility to enter into their work by learning about them, reading their reports, and praying for them. As I pray for those cells on the front lines, I sense their pain and struggle. We can all keep contact with members of the Body overseas and use them as our own personal touch corpuscles.

*

The world's needs are increasing like molten lava in an overdue volcano. Each day we watch a litany of news reports of famines, wars and epidemics, often even as we are deftly shovelling in the abundance of our own food. We casually flip past the advertisement appeals picturing babies with stomachs bloated by malnutrition. The needs are so overwhelming that, instead of shocking us to action, they make us callous, insensitive.

In some ways we are acquiring an intolerable burden of guilt that could immobilize us. Again, I think back to the ministry of Jesus. He healed people, but in a localized area. In His lifetime He did not affect the Celts or the Chinese or the Aztecs. Rather, He set in motion a Christian mission which was to spread throughout the world, responding to human needs everywhere. We must begin with our resources, our neighbourhood, our theatre of service. Although we cannot change the whole world individually, together we can fulfil God's command to fill the earth with His presence and love. When we stretch out our hand to help, we stretch out the hand of Christ's Body.

19

CONFRONTING

No one has greater love than the one who lays down his life for his friends.

JESUS

I would be remiss if I left the impression that skin's only functions are to inform us of our environment and relate to it by touch and appearance. Nature is never so lavish. Skin exists chiefly as a barrier, a Maginot line that keeps the inside in and the outside out. Without it some of our body parts would slosh around like Jell-O and we would lose our definition as an organism.

If I had to choose skin's most crucial contribution, I might opt for waterproofing. Sixty per cent of the body consists of fluids, and these would soon evaporate without the moist, sheltered world provided by skin. Or without skin, a warm bath would kill: fluids would rush in like water over a flooded spillway, swelling the body with liquid, diluting the blood, and waterlogging the lungs. Skin's tight barrier of shingled cells fends off such disasters.

Modern civilization taxes skin's capacities. We scrub with harsh detergents and soaps (which, ironically, may alter skin's acidic base and promote bacterial growth). On any Saturday we can subject our skin to the abuse of swimming in a chlorinated pool, spilling kerosene on our hands as a barbecue fire is lighted, cleaning paint brushes with turpentine and scouring it all off with abrasive powder and a roughened pad. But somehow skin survives.

Skin also offers a front-line defence against the hordes of bacteria and yeast that pepper its surface. Lennart Nilsson's superb microscopic photos of the body's surface reveal the tiny perforations of sweat pores and oil ducts as mammoth, jagged caverns providing entry into the deeper parts of the dermis. On the lips of these caves lurk glowing green bacteria and wildly spreading yeast. A single bacterium, which only lives for twenty minutes or so, can reproduce to a million in eight hours. Each one of us carries as many of these creatures on the surface of our bodies as there are people inhabiting this planet. Skin draws from an array of chemicals, electro-negative charges, and bands of defending cells to keep the marauders at bay.

Larger animals crawl in the fissures, too. Until this century in developed countries, mites, fleas, bedbugs, and lice were an accepted part of the skin's landscape. Thomas à Becket's hair shirt was studded with wriggling lice; Samuel Pepys had to return a wig that came from the hairdresser full of nits. The gentry of France, always concerned about proper behaviour, frowned on cracking fleas between the nails in public unless the gathering was a group of intimate friends.

Even today an eight-legged creature just a third of a millimetre long, the *Demodex folliculorum*, squirrels its way inside hair follicles and contentedly lives out its days in its burrow of choice, the eyelash. This mite, cigar-shaped and seemingly harmless, is found on almost every human examined. Male and female Demodexes merrily mate in the tunnel beside the hair, and as many as twenty-five of the creatures can congregate in one warm, oily fat gland.

Skin must also counter attacks by larger creatures, such

as Portuguese men-of-war, scorpions, ticks, fleas, blister beetles, and biting flies. Some bugs, thirsty for human juices, rush to constricted parts of the body where pressure squeezes the skin close to underlying blood vessels. Thus a chigger scampers across the body at a speed of three inches a minute until he reaches the constriction of underwear elastic. Ah, the epidermis is so thin and inviting there; he gorges himself on blood.

Large blows such as poundings and bruisings spread their impact across thousands of skin cells which spring back like a trampoline to absorb forces that could irreparably damage hidden organs.

It is a rough world out there, and the epidermis provides a continuous rain of sacrificed cells. This outer, horny layer is poised like curling cornflakes, ready to scale off and make room for moist, fresh cells from underneath. People who count such things estimate we lose ten thousand-million skin cells a day. Just shaking hands or turning a doorknob can produce a shower of several thousand skin cells; one trembles to calculate the effect of a game of tennis.

Dead cells linger on the surface of an arm which has been covered by a plaster cast for several weeks. But where do all the rest go? Pools of skin collect underneath sheets and some is lost to the breeze, but much stays around home. Up to 90 per cent of all household dust consists of dead skin – friendly scrapings of you, your family, your guests, waiting to be smoothed together with a soft cloth and shaken outdoors without a moment of gratefulness for the sacrifice represented. Replacement cells will grow back mainly between 12:00 p.m. and 4:00 a.m. while much of the body rests.

*

Once I was consulted by an enthusiastic young student just learning to play the guitar. With lines of worry in his face, he asked me to examine his fingertips; they were red and swollen and sometimes would bleed when he played. "Are they too weak to play – will I never be a guitarist?" he asked plaintively.

I had to laugh at the way he had been inspecting his own skin cells. Though they were part of him, working loyally on his behalf, he viewed them as a manager would his employees, wondering whether they were really contributing. I advised him to slow down. His skin was working furiously to keep pace with intense new pressures that scraped off the paper-thin epidermal layer before new cells could be marshalled. Soon the multiplication rate of his cells would catch up and layer his fingertips with hard callouses.

Of all the organs, skin seems to me the most sacrificial; it is no wonder one-fourth of a general practitioner's patients come because of skin ailments. Skin absorbs incredible abuse to maintain the equipoise of vital organs inside, which cannot tolerate a changing environment. A temperature increase of just seven or eight degrees would kill the whole body; thus, the skin is called on to act as a radiator, rushing fluids to the surface to evaporate and cool the body. Increased blood supply to the skin's surface dissipates the heat. On a summer day, as much as two gallons of perspiration may be shed to cool the active body.

In one sense, because the whole community of followers represents Christ's Body to the world, all Christians participate in the appearance function of skin. At times each of us also encounters the friction of being the advance guard of Christ's Body to the world. Yet I believe that just as our bodies must protect the delicate cells in the eye or the liver from harsh realities of the external environment, so the church includes individuals who need to be sequestered and allowed times of quiet contemplation. Others need protection during particularly vulnerable periods of their lives.

On behalf of all these, some members of the Body of Christ take up the front line, the exposed positions, and endure the trauma for the rest of us.

The skin is not a place for beginners. It is an advanced organ, programmed with the body's immunity and disease-fighting system. Allergies, smallpox, and tuberculosis are tested on its surface because it can represent the internal parts of the body and protect them. Christians, anxious to "put on a good front" for a watching world, eagerly push new converts to become the visible organ. Many are not wise or mature enough to handle the shock. I could rattle off a list of sports heroes who started out as featured speakers on the Christian athlete circuit and who fell away and today have no interest in Christian matters. They remind me of the tender, swollen cells of the young guitar player, still not adjusting to the increased stress of being pulled across steel wire.

New converts, especially susceptible to the dangers of their alien environment, need protection to learn the ways of the body. If the apostle Paul needed a lengthy time of reflection, should not we ask the same for new Christians today?

*

Not all of us will be called to the front lines. And those involved in more humble service inside the Body face their own unique danger: they feel inferior to the parts of the Body that are more visible. Can typing or cleaning hospital rooms contribute to the kingdom in the same way as the activities of the visible representatives of the faith? The Bible often focuses on those rare people who were called on to lead the way and forge new territory for religious faith and practice. They are important models for us, to be sure. But we will not all be apostles, and there is no hint in the Bible that we should be. On the whole, the church is populated by ordinary citizens who are different mainly because of the allegiance of their lives.

Some are called to the front, such as Mother Teresa, Corrie ten Boom, and Billy Graham. From us they deserve support and prayer, not envy, for life on the surface of the Body is never easy.

The history of the church is dotted with cells who were willing to live at the touch-point of friction; these men and women did not shrink from bruisings or scorching heat or unbearable stresses. I read the list of heroes in Hebrews 11 as a roll call of martyrs who fought on the front lines, ''Who shut the mouths of lions, quenched the fury of the flames, and escaped the edge of the sword; whose weakness was turned to strength; and who became powerful in battle and routed foreign armies. Women received back their dead, raised to life again. Others were tortured and refused to be released, so that they might gain a better resurrection. Some faced jeers and flogging, while still others were chained and put in prison. They were stoned; they were sawed in two; they were put to death by the sword. They went about in sheepskins and goatskins, destitute, persecuted and mistreated – the world was not worthy of them. They wandered in deserts and mountains, and in caves and holes in the ground'' (Hebrews 11:33–38).

Today, Christians under oppressive regimes are being persecuted for their faith. Alexander Solzhenitsyn reminds us of the great reservoir of suffering that has accumulated among Russian Christians and the legacy they have bequeathed to the world.

I think of my own mother, from a society home in suburban London, who went to India as a missionary. When Granny Brand reached sixty-nine she was told by her mission to retire, and she did . . . until she found a new range of mountains where no missionary had ever visited. Without her mission's support she climbed those mountains, built a little wooden shack, and worked another twenty-six years. Because of a broken hip and creeping paralysis she could only walk with the aid of two bamboo sticks, but on the back of an old horse she rode all over the mountains, a medicine box strapped behind her. She sought out the unwanted and the unlovely, the sick, the maimed and the blind, and brought treatment to them. When she came to settlements who knew her, a great crowd of people would burst out to greet her.

My mother died in 1974 at the age of ninety-five. Poor nutrition and failing health had swollen her joints and made her gaunt and fragile. She had stopped caring about her personal appearance long ago, even refusing to look in a mirror lest she see the effects of her gruelling life. She was part of the advance guard, the front line presenting God's love to deprived people.

*

Another woman, also serving on the front lines, captures for me in a single image all the elements of the skin of Christ's Body. I visited a nun, Dr Pfau, in the 1950s outside Karachi, Pakistan, in the worst human squalor I have ever encountered. Long before I reached her place, a putrid smell burned my nostrils. It was a smell you could almost lean on.

Soon I could see an immense garbage dump by the sea, the accumulated refuse of a large city that had been stagnating and rotting for many months. The air was humming with flies. At last I could make out human figures – people covered with sores – crawling over the mounds of garbage. They had leprosy, and more than a hundred of them, banished from Karachi, had set up home in this dump. Sheets of corrugated iron marked off shelters, and a single dripping tap in the centre of the dump provided their only source of water.*

But there, beside this awful place, I saw a neat wooden clinic in which I found Dr Pfau. She proudly showed me her orderly shelves and files of beautifully kept records on each patient in the dump. The stark contrast between the horrible scene outside and the oasis of love and concern inside her tidy clinic burned deep into my mind. Dr Pfau was daily exhibiting all the properties of skin: beauty, sensitivity to needs, compliancy, and the steady, fearless application of divine love through human touch. All over the world people like her are fulfilling Christ's command to fill the earth with His presence.

* Today the garbage dump is gone, and Dr Pfau serves as a senior leprosy doctor in Pakistan at a modern hospital.

MOTION

20
MOVEMENT

*In the absence of any other proof, the thumb alone would
convince me of God's existence.*

ISAAC NEWTON

A kindly looking old gentleman with a more-than-
prominent nose and a face seamed with wrinkles crosses
the stage. His shoulders slump and his eyes seem sunken
and cloudy – he is over ninety years old. He sits on a stark
black bench, adjusting it slightly. After a deep breath, he
raises his hands. Trembling slightly, they poise for a
moment above a black and white keyboard. And then the
music begins. All images of age and frailty slip quietly from
the minds of the four thousand people gathered to hear
Arthur Rubinstein.

His programme tonight is simple: Schubert's *Impromptus*,
several Rachmaninoff Preludes, and Beethoven's familiar
Moonlight Sonata, any of which could be heard at a music
school recital. But they could not be heard as played by
Rubinstein. Defying mortality, he weds a flawless technique
to a poetic style, rendering interpretations that evoke
prolonged shouts of "Bravo!" from the wildly cheering
audience. Rubinstein bows slightly, folds those marvellous
nonagenarian hands, and pads offstage.

I must confess that a bravura performance such as that

by Rubinstein engrosses my eyes as much as my ears. Hands are my profession; I have studied them all my life. A piano performance is a ballet of fingers, a glorious flourish of ligaments and joints, tendons, nerves, and muscles. I must sit near the stage to watch their movements.

From my own careful calculations I know that some of the movements required, such as the powerful arpeggios in the *Moonlight*'s third movement, are simply too fast for the body to accomplish consciously. Nerve impulses do not travel with enough speed for the brain to sort out that the third finger has just lifted in time to order the fourth finger to strike the next key. Months of practice must pattern the brain to treat the movements as subconscious reflex actions – "finger memory" musicians call it.

I marvel too at the slow, lilting passages. A good pianist controls his or her fingers independently, so that when striking a two-handed chord of eight notes, each of the fingers exerts a slightly different pressure for emphasis, with the melody note ringing loudest. The effect of a few grammes more or less pressure in a crucial pianissimo passage is so minuscule only a sophisticated laboratory could measure it. But the human ear contains just such a laboratory, and musicians like Rubinstein gain acclaim because discriminating listeners can savour their subtlest nuances of control.

*

Often I have stood before a group of medical students or surgeons to analyse the motion of one finger. I hold before them a dissected cadaver hand, almost obscene-looking when severed from the body and trailing strands of sinew. I announce that I will move the tip of the little finger. To do so, I must place the cadaver hand on a table and spend perhaps four minutes sorting through the intricate network of tendons and muscles. (To allow dexterity and slimness for actions such as piano playing, the finger has no muscles in itself; tendons transfer force from muscles in the forearm and palm.) Finally, when I have arranged at least a dozen muscles in the correct configuration and tension, with a delicate movement I can manoeuvre them so the little finger firmly moves without the proximal joints buckling.

Seventy separate muscles contribute to hand movements. I could fill a room with surgery manuals suggesting various ways to repair hands that have been injured. But in forty years of study I have never read a technique that has succeeded in improving a normal, healthy hand.

I remember my lectures as I sit in concert halls watching slender fingers pump up and down or glissade across the keyboard. I revere the hand; Rubinstein takes its function for granted. Hands are his obedient servants; often he closes his eyes or gazes straight ahead and does not even watch them. He is not thinking about his little finger; he is contemplating Beethoven and Rachmaninoff.

Scores of other muscles line up as willing reinforcements for Rubinstein's hands. His upper arms stay tense, and his elbows bend at nearly a ninety-degree angle to match the keyboard height. Shoulder muscles rippling across his back must contract to hold his upper arms in place, and muscles in his neck and chest stabilize his shoulders. When he comes to a particularly strenuous portion of music, his entire torso and leg muscles go rigid, forming a firm base to allow the arms leverage. Without these anchoring muscles, Rubinstein would topple over every time he shifted forward to touch the keyboard.

*

In order to observe the types of artificial hands that scientists and engineers have developed through years of research and millions of dollars of technology, I have visited facilities that produce radioactive materials. With great pride scientists demonstrate their skilled machines that allow them to avoid exposure to radiation. By adjusting knobs and levers they can control an artificial hand whose wrist supinates and revolves. Recent models even possess an opposable thumb, an advanced feature reserved for primates in nature. (Only we humans, though, can join the tips of our index fingers with our thumbs, allowing us to grasp, retain, and handle objects easily and precisely.) Smiling like a proud father, the scientist wiggles the mechanical thumb for me.

I nod approvingly and compliment him on the wide range of activity the mechanical hand can perform. But he knows, as I do, that compared to a human thumb his atomic-age hand is clumsy and limited, even pathetic – a child's Play Doh sculpture compared to a Michelangelo masterpiece. A Rubinstein concert proves that.

*

Six hundred muscles, which comprise 40 per cent of our weight (twice as much as bones), burn up much of the energy we ingest as food in order to produce all our movements. Tiny muscles govern the light permitted into the eye. Muscles barely an inch long allow for a spectrum of subtle expression in the face – a bridge partner or a SALT negotiator learns to read them as important signals. Another, much larger muscle, the diaphragm, controls coughing, breathing, sneezing, laughing, and sighing. Massive muscles in the buttocks and thighs equip the body for a lifetime of walking. Without muscles, bones would collapse in a heap, joints would slip apart, and movement would cease.

Human muscles are divided into three types: smooth

muscles control the automatic processes which rumble along without our conscious attention; striated muscles allow voluntary movements, such as piano-playing; and cardiac muscles are specialized enough to merit their own category. (A hummingbird heart weighs a fraction of an ounce but beats eight-hundred times a minute; a whale heart weighs one-thousand pounds – in contrast to either, the human heart seems dully functional, but does its job well enough to get most of us through seventy years with no time off for rest.)

Surrounded as we are by man-made motion – aeroplanes, dune buggies, colour dots dancing across a TV screen – we can grow numb to the sheer exaltation of movement made possible by muscles. But even lower forms of animal life display impressive feats. A common housefly's muscles respond in one-thousandth of a second, which is why not many are caught with the bare hand. The despised flea performs acrobatic leaps and somersaults, which, if factored up to human size, would make our best Olympians quit in dismay. Visit a zoo with an underwater window and watch the seals and sea lions, awkward and ponderous on land, infuse the word "graceful" with new meaning. Stand in a farmyard and watch a swooping swallow redefine flight.

As is often the case, the human being has a more conservative, scaled-down range of movement. We cannot see like an eagle, hear like an owl, or glow like a firefly, nor can we run like a dog, leap like a gnat, or fly like a goose. But we do have enough potential packed in our muscles to allow the Bolshoi Ballet and the sports of ice skating and gymnastics. On TV the performers are models of weightless beauty, gliding through the air, pirouetting on a single toe, dismounting from a high bar with a light spring. But in person, close to the event, the grace is seen as a by-product of hard work. It is *noisy* there, all shocks and thuds and creaking boards and panting, sweating bodies. That humans can transform such strenuous muscular activity into fluidity and grace is a tribute to the dual nature of motion: robust strength and masterful control.

21

BALANCE

Christianity got over the difficulty of combining furious opposites, by keeping them both, and keeping them both furious.

G. K. CHESTERTON

The movements of Rubinstein or Baryshnikov or Heiden do not come cheaply. The motor cortex of the brain, on which will be written all the coding for intentional movement, starts out blank as a washed chalkboard. Although the seeds of instinctive behaviour are there, an infant, dominated by gravity, cannot hold his head or trunk upright. His hand and leg movements are abrupt and jerky, as in an old silent movie. He learns fast, however, lifting his head in one month and his chest in two. In seven months, if all goes well, he sits upright without support. At age eight months the infant can stand unassisted, but on the average it takes seven more months for him to walk smoothly at the speed of one footstep per second, without consciously thinking of the action.

If we traced all the body signals involved in walking, we would find in that grinning, perilously balanced toddler a machine of unfathomable complexity. Over one hundred million sense cells in each eye compose a picture of the table he is walking toward. Stretch receptors in the neck relate the attitude of his head to the trunk and maintain appropriate muscle tension. Joint receptors fire off messages that report the angles of limb bones. The sense organs inside the ear inform the brain of the direction of gravity and the body's balance. Pressure from the ground on each toe triggers messages about the type of surface on which he is walking.

Just for the toddler to stand, the muscles which oppose each other in the hip, knee, and ankle must exert an equal and opposite tension, stabilizing the joints and preventing them from folding up. "Muscle tone" describes the complex set of interactions that keeps all the infant's muscles in a mild state of contraction, making his erect posture as active and strenuous as the movements that follow it.

A casual glance down to avoid a toy on the carpet will cause all these sense organs to shift dramatically: the image of the ground moves rapidly across the retina, but the inner ear and stretch receptors assure the brain the body is not falling. Any movement of the head alters the body's centre of gravity, affecting the tension in each of the limb muscles. The toddler's body crackles with millions of messages informing his brain and giving directions to perform the extraordinary feat of walking.

*

Muscles rely on an advanced hierarchy to organize the individual cells. Muscle cells – long, sleek bodies with dark nuclei – perform just one action: they contract. They can only pull, not push, as two protein molecules interact and the molecules slide together like the teeth of two facing combs. Cells unit in strands called fibres, resembling coils of rope, and fibres report to a further hierarchy called a motor unit group.

One motor nerve controls a motor unit group, wrapping

its end plates around the muscle group as an octopus would encircle a pole. When that nerve gives a signal, all of its muscle fibres immediately become shorter and fatter. Some fibres are "fast-twitch" for short bursts of energy while others, "long-twitch," are less quickly fatigued. Muscles fibres adhere to what is called the "all or none" principle. They do not have a variable throttle of energy, but a simple on-off switch. Variations of strength, as when Rubinstein lightly taps a key or pounds it mightily, occur because of the quantity of motor units firing off at any moment.

Conductors of large choirs warn their singers not to take breaths at the end of a pianissimo measure, since the sound of many singers inhaling would be audibly distracting. Rather, they should gasp for air in the middle of a measure, staggering their breathing so that the large choir continues singing while just a few members inhale at any one instant. Unlike a choir, however, a muscle cannot ask its members to sing softly. To vary the volume the biceps simply alter the number of participants. Each motor unit takes a rest when needed, but the muscle's contraction stays steady.

Rarely will all the motor units in a large muscle fire simultaneously. Occasionally, adrenaline induces feats of great strength, called hysterical strength, such as a mother lifting a car off her child — perhaps then all the motor units are galvanized into action.

The muscle "choir" can be literally heard if a needle is inserted into a muscle and attached to a machine that transforms electrical energy into sound. Click-click-click: a constant stream of messages reports the activity of muscle tone. Slowly flex the biceps, and the volley of clicks accelerates. Move the arm abruptly, and the clicks crescendo to machine-gun frequency. The cells never stop clicking, and they adjust instantly, within fractions of a second, when the brain calls for sudden activity.

As the meter records the stream of static flowing from just one muscle area the size of a needle point, hundreds of other muscles go wholly undetected. A large and crucial group of them fire off whether or not we think about them: the automatic muscles controlling our eyelids, breathing, heartbeat, and digestion. It is as if the wisdom of the body does not trust the forgetful, erratic free will with these life-or-death functions. So protected are they that we cannot voluntarily stop our heartbeat or breathing. No one can commit suicide by holding his breath; accumulating carbon dioxide in the lungs will trigger a mechanism to override conscious desire and force the muscles of ribs, diaphragm, and lungs to move.

Consider the electrical network linking every home and building in metropolitan New York City. At any given second, lights are turned on and off, toasters pop up, microwave ovens begin their digital countdowns, water pumps lunge into motion. Yet that enormous interlinking of decisions and activities is marked by randomness. A far more complex switching system is operating in your body at this second as you read this book, and it is perfectly controlled and orderly. When you reach the end of this page, you will turn it with your fingers, still only vaguely aware of the complex systems that allow such an act.

*

In the physical body as well as the spiritual, a muscle must be exercised to continue growing. If, through paralysis, we lose movement, atrophy will set in and muscles will shrink away until they are absorbed by the rest of the body. Similarly, Christ's Body shows its health best by acting in love toward other human beings. When it cuts back on active response to pain and injustice, it begins to waste away and weaken. If an organic faculty is not used, it will degenerate; parasites pitifully exhibit this law of nature.

One aspect of motion in Christ's Body continues to puzzle me, though. Even when it is being exercised in history, it seems marked by a confused, convulsive nature. Pick any century, and the history of the church then will include splits and divisions, heated debates about the role of social concern, and sadly excessive reactions to non-Christian influences. Because church history includes these tentative, spastic flailings of activity, we easily discount the effectiveness of the Body's motion.*

As I look closer at the biology of motion, though, I can better grasp how seemingly disconnected spurts of energy can actually contribute to fluidity. In the human body a motion does not result from all parts unanimously contributing the same activity; in fact, every action has an equal and opposite reaction. We have seen that muscles are paired antagonistically so that when the triceps contracts the biceps relaxes, and vice versa. But one of the pioneers of neurophysiology, Sir Charles Sherrington, demonstrated that *all* muscular activity involves inhibition as well as excitation. In every muscular sentence there is a balancing "but".

The knee jerk, which involves only two muscles, illustrates Sherrington's principle. When a doctor taps a patient's knee, the muscle on the front of the thigh springs into action, excited. But the action is not possible unless the back of the thigh, which bends the knee, is actively inhibited and chooses not to contract. Two stimuli are equally powerful; one leads to action, one to inaction. In complex movements, like walking or hitting a baseball,

* At least a part of the confusion stems from the fact that the visible, organized church may be very different at any given time from the true church – the Body of Christ. A pastor or even a bishop may at some moment of history have been outside of the Body and working against it.

hundreds of opposing reactions occur simultaneously. All muscular action, therefore, involves this policy of give-and-take. Sherrington expounded on this concept: "It has been remarked that Life's aim is an act, not a thought. Today the dictum must be modified to admit that, often, to refrain from an act is no less an act than to commit one, because inhibition is coequally with excitation a nervous activity."[1] Not to decide to act is to decide.

A harmony of inhibitions synchronizes the whole body, coordinating heartbeats with breathing and breathing with swallowing, setting muscle tone, adjusting to all changes in movement. In short, inhibition keeps one part of the machine out of the way of the other.

This biological principle may help explain what at first glance appears as a troubling recurrence in the history of the church. The Body of Christ has moved by extreme, exaggerated reflexes. On the very issue of activity versus inactivity, debate broke out in the early years of the church. In behaviour, as Charles Williams has pointed out, there are two opposite tendencies. "The first is to say: 'Everything matters infinitely.' The second is to say: 'No doubt that is true. But mere sanity demands that we should not treat everything as mattering all that much.' "[2] The rigorous view leads to a sharpened, intense view of the world that sees all actions as having eternal consequences. In its worst forms it can evolve into pharisaic legalism and the intolerance of "holy" crusades. The relaxed view, contributing sanity, can at its worst drift toward inactivity, a "who cares?" attitude toward injustice and sin.

The apostle Paul, notably in Galatians and Romans, fought a pitched battle against both extremes, on the one hand excoriating legalists for perverting God's grace and on the other hand upholding Christian works as a normal outgrowth of new life.

In relating to the larger world, too, Christians have oscillated between opposing forces. In the first two Christian centuries, the Way of Affirmation and the Way of Negation sprang up, each attracting ardent followers. The Way of Affirmation established strict church policy: "If any bishop or priest or deacon, or any cleric whatsoever, shall refrain from marriage and from meat and from wine . . . let him be either corrected or deposed and turned out of the Church."[3]

Abstainers from marriage and feasting were labelled "blasphemers against creation," and the affirmers had many targets, what with all the athletes of God running around thin and naked in the desert. The paradox was hardly new: Jesus pointed out that John the Baptist had been blasted for his asceticism while He, the Son of God, was gossiped about as a winebibber and glutton (Matthew 11:19). Each tendency extracted something good from the conflict: The Way of Affirmation bequeathed us great art and romantic love and philosophy and social justice while the Way of Negation contributed the profound documents of mysticism that could only come from undisturbed contemplation of the holy.

Christians today are trying to balance the church's aesthetic appreciation against critical needs in an increasingly overpopulated world. Is it possible to maintain beautiful art and lavish architecture in view of the resources such activities consume? Some are rediscovering the need for community, which, in a society as stubbornly democratic as the West, may need a highly structured form. Mission leaders constantly struggle with the tensions created between their twin goals of ministernig spiritually and materially through evangelism and social concern. Even a tiny congregation may reflect the counterbalancing tendencies.

If I visit a Christian community of young political radicals who strongly oppose the American government and advocate total pacifism and intentional poverty, I may come away with a distorted view of what Christian activity in the world should look like. Yet such a Christian counterculture can, by the process of inhibition, temper the activity of the institutionalized church, smoothing out its insensitive movements, calling it back to a radical awareness of justice. Perhaps their contribution can keep the body from toppling over to one side.

The unifying factor in such debates must be a common commitment to the Head, Jesus Christ. We will disagree on interpretations of what He said and meant and what is the best means of accomplishing those goals in a hostile society. But if we fail to find fellowship in our mutual obedience to Him, our actions will be seen not as reciprocal, counterbalancing forces necessary for movement, but as spastic, futile contractions.

22

DYSFUNCTIONS

Our faculties are like those smelting works that can only take ore of a high degree of impurity; when the light is too bright we cannot see.

MALCOLM MUGGERIDGE

A man entered my office in India, a blubbering hulk of a man. He was a successful Australian engineer who had worked in India for many years. But his neck twitched so violently that every few seconds his chin smashed into his right shoulder. He had spasmodic torticollis, or twisted neck syndrome, a peculiarly debilitating condition usually caused by a deeply rooted psychological disorder.

Between the spastic flingings of his chin, my patient described his despair. To compound his reasons for self-pity, he was short and fat and had a history of alcoholism. The torticollis, he said, had begun soon after a visit to Australia. A confirmed bachelor throughout his time in India, he had returned from Australia with a wife – a gorgeous woman, taller and younger than he, who immediately became the object of much village gossip. What had she seen in him? What had prompted such a mismatch?

I referred the engineer to a psychiatrist, for I could do nothing except sedate him temporarily. The psychiatrist confided to me his suspicion that the engineer's condition had developed out of his anxiety over not being able to measure up to his new wife. He offered a diagnosis, but no hint of a cure. The engineer returned to me in a few weeks, even more weighted down with despair. Slovenly kept, with his neck wrenching spasmodically, he was an object of great pity.

When he sat alone, unnoticed by anyone, his neck rarely contorted. But as soon as someone began a conversation with him, his chin would shoot over to his shoulder, aggravating a chronic, spongy bruise. I researched the condition and worked with him, but nothing helped other than sedation and the temporary relief that followed an injection of his nerve roots with Novocaine. Finally he reached the point of utter despondency and attempted suicide. He insisted, with a firm and resolute edge to his voice, that he would try again and again until he succeeded. He could no longer continue living with his anarchic neck.

I tried to send him abroad, since there was no neurosurgeon in India, but he refused. Reluctantly I agreed to attempt a dangerous and complicated operation that involved exposing his spinal cord and the base of his brain. I had never tried a procedure quite so complex, but the man insisted his only alternative was suicide.

I cannot recall an operation plagued with as many mishaps as that one. We had improvised an extension to a regular operating table so that the patient could lie face down, as on a neurological table. Unfortunately, this made it difficult for the anaesthetist to replace the tube in his trachea when it became dislodged. The resulting poor oxygenation made bleeding profuse, and the cautery short-circuited at the critical time when we most needed it to control bleeding. Then all the hospital lights failed, and I was left with only a hand-held flashlight and no cautery just when the spinal cord was coming into view. To add to the stress, I had neglected to empty my bladder before surgery and was most uncomfortable throughout.

Between these distractions, I tried to concentrate on some very delicate cutting. After exposing the spinal cord and lower brain, I traced the hair-like nerves that supplied the spastic muscles in his neck. Any slight quiver of the scalpel could have cut a bundle of nerves, destroying movement or sensation.

Somehow, in spite of these difficulties, the surgery proved successful. When the engineer awoke, his back humped with a bandage, he discovered that the feared neck movement no longer plagued him. It couldn't, of course, for I had cut the motor nerves which led from the spinal cord to the muscles that turned his neck; he could not make the movement that had previously dominated him. That group of muscles had been totally rejected because of their rebellion against orders from the brain. Gradually, through lack of use, they were absorbed into the body.

*

When people see someone with a spastic muscle, they often assume the muscle itself is malfunctioning. Actually, the muscle is perfectly healthy, not diseased. In fact, it is well-developed because of frequent use. The malfunction stems from the muscle's relationship to the rest of the body; it demonstrates its strength at the wrong times, when the body neither needs nor wants it to function. A spastic muscle may, as in the case of the Australian engineer, cause embarrassment, pain, and deep despair.

Just as aberrant fat cells can lead to a harmful tumour by hoarding the body's resources, so spastic muscles can interfere with the body's normal movement. Quite simply, a spastic muscle disregards the needs of the rest of the body; its dysfunction is closer to rebellion than disease.

Acts of love – healing, feeding, educating, proclaiming Christ – are the spiritual Body's proper functions of movement. Even these motions, though, which appear wholly good, can fall prey to a dangerous dysfunction. Like the spastic muscle, we can tend to perform acts of kindness for our own benefit, for our sakes and reputations. In ministering to physical and emotional needs, we are especially susceptible to the temptations of "playing God" and self-contented pride. Love, having become a god, seeks to become a demon. Those of us in Christian work, I have found, consistently come against this subtle tendency toward pride. Someone comes to me for spiritual counsel, and I give it. But before they have walked out of my room I'm congratulating myself on what a fine counsellor I am.

Jesus' disciples, the first trainees to represent Him, consistently stumbled at this point. They argued about such petty issues as who was the greatest disciple and who would have the greatest honour in heaven (Matthew 20:20–23). Jesus lectured them on the need for self-sacrifice, pulled children from the crowd to picture the meek attitude they should have, even washed their dirty feet to illustrate service. It did not seem to sink in – not until after that dark day on Calvary.

*

I have no desire to make judgments or name Christians today who seem to be exercising their muscles in a self-serving rather than a Body-serving way. But I do wonder about the explosive growth of the electronic church. This powerful new muscle can reach millions of people and also collect millions of dollars in revenue. But does the medium give some leaders too much leverage and power? As a former missionary in a helping role, I know too well the human weaknesses that lead to spiritual pride. Media evangelists and Christian speakers and performers have described to me their unique pressures. They can easily bask in the glow of warm acceptance and ego-satisfying comments from adulating fans. Executives in Christian corporations and pastors are subject to the same temptations of pride and status.

None of us is exempt. Radical Christians who urge action in the inner city, politically conservative Christians who give large sums of their investments to missions, seminary students who glory in their new-found knowledge, church members who fill out committees within the church – all of us need to come back to the image of the Son of God kneeling on a hard floor and unbuckling sandals covered with choking Palestine dust. We cannot find real fulfilment by demonstrating individual strength as a muscle unit in Christ's Body. Rather, our activity must be for the sake of the Body. If we loyally serve Christ, and applause or even fame results, we will need special grace to handle it. But if we consciously seek applause or fame or wealth, for whatever end result, the effect will be like the spastic contraction of a once-healthy muscle. Like Ananias and Sapphira, we will have turned a good act into an impure act because of our impure motive.

*

Movement in the Body, then, requires the smooth and willing cooperation of many parts who gladly submit their own strength to the will of the Head. If they act apart from the Head's orders, their action, though powerful and impressive, will not benefit the body.

Motion also involves another severe problem which can cripple the body. When parts work together closely, they generate friction. I was reminded of this danger when a famous pianist in England consulted me. She said a specific pain was interrupting all of her performing. No longer could she concentrate on the flow of music or the rhythm. Instead her mind focused on the pain that would shoot through her hand whenever the thumb moved at a certain angle to her wrist. She had recently cancelled a series of concerts because of that grating pain, even though all her other skills – music interpretation; muscle action, sense of touch, and timing – were intact.

I told her that the trouble emanated from a small, rough arthritic area between the two wrist bones at the base of her thumb. I suggested she continue to play but try to move that joint minimally. "But how can I think about Chopin when I have to worry about the angle of my thumb?" she protested. Each time she started to play, her attention riveted on the painful friction of that one roughened little joint.

Treating patients such as this pianist prompted me to study the type of lubrication our joints use and I learned that one of the most astonishing things about our bodies is how our joints ordinarily work so smoothly and free of pain. At the Cavendish laboratory in Cambridge, England, a team of chemists and engineers compared the frictional properties of the cartilage lining our joints with that of materials the engineers use for bearings. They were seeking a material suitable for use in artificial hips. Initially, they calculated that the friction present in the knee of an ox was one-fifth that of highly polished metal – about the same friction as ice on ice. It did not seem possible that biology could offer a joint five times more efficient than anything engineering science had ever achieved.

They researched further and found that joint cartilage is filled with tiny channels full of a synovial fluid. The cartilage is compressible, and as a joint moves, the part of the cartilage bearing the strain compresses, causing jets of fluid to squirt out from these canaliculi. The fluid forms a sort of forced pressure-lubrication which lifts the two surfaces apart. When the joint moves further, a different part of the surface bears the stress; fluid in the new area squirts out while the area just relieved of pressure expands and sucks its fluid in. Thus in active movement the joint surfaces do not really touch, but float on jets of fluid. The engineers were astonished, for boundary lubrication and pressure lubrication were recent developments in engineering – they had thought.

In the Body of Christ, joints are those special areas of potential friction where people work together in some stressful movement. In a body at rest there is little need for resistance against friction, but as soon as muscles and bones start producing activity, joints become critical attention points. Considering how soon joints and bearings need attention in a new machine, my joints amaze me with their ability to last for decades without squeaking or grinding. But despite their remarkable powers of lubrication, joints can break down as their gliding surfaces are injured or start to wear thin.

Quite commonly in old age friction will begin to cause joints to ache and throb – a natural response to years of wear. In Christ's Body, this natural wear is sometimes seen in the intolerant way older, wiser Christians may judge those who have a new enthusiasm for the faith but much to learn about behaviour or doctrine. In recent years the church has absorbed a large influx of new people, especially the "Jesus people" of the sixties and waves of charismatic Christians in the seventies. Some older Christians have found themselves getting irritable and intolerant in their relationships with these new members. Sometimes the grace of God must come in the form of little squirts of synovial fluid that help the old to understand and get along with the young and that help the young to understand what it must be like to have thin cartilage.

*

Far more serious than this natural ailment of the joints is the condition of rheumatoid arthritis, which may cripple even the young. We really do not know the cause of this disease which somehow produces a hypersensitivity in the cells of the joints. Suddenly a joint becomes flooded with enzymes that normally occur only when bacteria and foreign protein call for defence mechanisms. A usually healthy reaction turns cannibalistic, and the cells of the synovial membrane respond as if they were inflamed by infection. When we open up the joints and examine them, we can find no enemies, just the angry presence of defensive cells vainly attacking the body's cartilage and ligaments. A dreadful civil war has broken out: the defence mechanism itself has become the disease.

Various theories attempt to explain rheumatoid arthritis. One proposes that there really is an enemy, but we have not isolated and identified it yet. Whatever theory is true, this overreaction causes painful, irreversible harm. Even if a real enemy is present, that enemy would most likely inflict less damage than these cells do by reacting against it.

Spiritual rheumatoid arthritis sometimes attacks the work of the Christian church. Members become hypersensitive, taking offence at imagined criticism. Their own dignity and position become more important than the harmony of the group. Or, they may choose a minor doctrinal issue and make agreement on it the determinant of spiritual unity.

The lesson here is so obvious that it hardly needs to be clarified; yet it does, certainly, need to be applied. Do friction and tension flare up in my environment? Could they result from my own righteous indignation against wrong within my family or colleagues or church? Could my anger be causing more harm than the wrong I am angry about?

Arthritis strikes at joints because there the friction caused by movement takes its toll. Some may think Christians are less susceptible to friction because of the ideals and goals they hold in common. But in fact Christian work can increase friction as the pressures to "be spiritual" compound normal working tensions. At the Christian Medical College in India we had a psychiatrist whose clients frequently were missionaries. Being highly motivated, working in lonely places, often with just one partner, missionaries seem to fall prey to acute personal tensions. Often they refuse to admit their problems until friction has undone all the good they have accomplished.

Two women will serve in one station together, with only each other for fellowship. While they face a tremendous task together, what cracks them is not the size of the task but the grating, everyday frictions of working together. And they won't express the tension because they believe that to be unchristian and they don't like to admit a real problem exists. So they bury it, channelling it into emotional and physical damage. When the frictions finally come out, they may root back to such trivial things as an ill-timed joke, a tendency to snore, or the way a roommate picks her teeth.

People sometimes assume the Christian life brings with it a natural immunity against friction, but it clearly does not. The human body goes to incredible lengths to prevent friction, and the Body of Christ should be as careful to lubricate possible conflicts as we move in common activity.

23
HIERARCHY

*The neuron is like a miniature person – having a personality,
having an array of unlike parts, having actions both
spontaneous and upon stimulation. . . . It speaks finally with
one voice, which integrates all that went before.*

THEODORE H. BULLOCK

I have casually referred to the linking force that races
through the body as electricity. Electricity? Today's
assumption was yesterday's wild adventure. The very word,
charged with lightning bolts and immolated bodies, was
as terrifyingly mysterious to former generations as atomic
energy is to ours. Today we manufacture electricity, with
local utility boards determining how many dollars should
be exchanged for the prompt conveyance of it to our homes.
But still a thousand jagged streaks of fire assault the earth
every minute in the form of lightning. Only a brave man
flaunts Zeus.

What relevance could that dreaded juice of the heavens
have to the billions of tiny nerve cells unifying me? Luigi
Galvani, an Italian who lived thirty years after courageous
Ben Franklin, launched his kite in the labyrinth of human
nerves. Before Galvani, every scientist and doctor since
A.D. 130 had faithfully followed the theory of the Greek
physician Galen, who elegantly described communications

in the body as an uninterrupted flow of ethereal "animal spirits" through a network of hollow tubes. His theory served the age well. What era but our own would attempt to reduce the tingle of a lover's desire, the surge of response to Vivaldi's music, and the holy mysticism of a saint to the quantifiable formulas of chemical actions and electrical impulses?

Galvani, poor soul, could not anticipate the reductionist lengths his discovery would lead to; he merely brought a few frogs home for dinner one cloudy day and hung them in his porch. Following one of those crazy, implausible hunches that have formed the history of science, he beheaded the frogs, skinned them, and ran a wire from a lightning rod to the frogs' exposed spinal cords. He recorded what happened next as a summer thunderstorm growled across the Bologna sky: "As the lightning broke out, at the same moment all the muscles fell into violent and multiple contractions, so that, just as does the splendour and flash of the lightning, so too did the muscular motions and contractions . . . precede the thunders and, as it were, warn of them."[1]

Galvani was a scientist; had he been a writer he would have described the anxious astonishment on the faces of his guests who watched beheaded frogs jerk and twitch as if they were kicking across a pond. Electricity in the atmosphere had flowed through the nerves of the frogs and stimulated movement in otherwise dead animals.

Galvani performed many other experiments on frogs, some of which have been so apocryphized over the years that it is hard to know what really did occur. A shy man, he published his findings relatively late in life, letting his nephew defend most of his theories in public. But his most consequential discovery came one bright day when he hung several beheaded frogs on copper hooks above the iron railing on his porch. Whenever one of the frog legs drifted toward the railing and made contact, it jerked violently.

Dead frogs jumping during a lightning storm are one thing, but high-kicking on the porch on a sunny day – that's the kind of discovery to set the scientific community on its ears. And so it did.

Galvani's rival, Alessandro Volta, concluded that the electric current had everything to do with two dissimilar metals joined by a conductor. He went on to invent the battery, and we have him to thank for exploding scoreboards, the electric newspaper on Times Square, the floodlit Wrigley building in Chicago, and a battery that starts a car on a below-zero morning.

Galvani stubbornly insisted the reaction came from "animal electricity," and we have him to thank for EKG monitors, bio-feedback machines, electric shock treatment, and untold millions of dead frogs' legs hopping madly in medical school laboratories.

Another century and a half would pass before body explorers finally came up with a reasonable explanation of how electricity travels through the body. Obviously, it couldn't flow like the current sputtering behind each wall receptacle, not over nerves so fine that a hair-width bundle of them contains 100,000 separate "wires." Rather, electric current inside us passes through the chemical interactions of sodium and potassium ions, and now medical textbooks portray colourful drawings of nerve cells with plus signs outside the membrane and minus signs inside, illustrating how molecules carry the nervous messages like runners passing a torch.

*

A cell called the neuron is the most important unit in communication inside the body. Twelve billion neurons are poised for action at birth. Every other cell in the body dies away and is replaced every few years, but not the neuron. How could we function if the reservoirs of memory and our information about the world were periodically sloughed away? When the neuron dies, it does not grow back. By unanimous decree of medical specialists, the neuron is declared the most significant and interesting cell in the entire organism.

Biology textbooks picture single neurons, plucked out of the body and stained in idealized form as they never appear in nature. But even from such caricatures one can sense the neuron's grandeur. It begins with a maze of incredibly thin, lacy extensions called dendrites which, like the root-hairs of a tree, funnel to a single shaft. On afferent neurons, which carry messages to the brain, these dendrites extend to whatever part of the body they are reporting stimulation from. On efferent neurons, which control muscles, branches wrap around muscle fibres, terminating in the endplates that directly control muscular activity.

The medical student who has studied acetate renderings of organs, neatly labelled and layered, is in for a rude shock the first time he or she opens a cadaver and finds a mess of bloody organs all looking approximately alike and nudging each other for room. Likewise, a surgeon never encounters a neuron standing in stark relief apart from the body. He sees hundreds, perhaps thousands, joined together in rope-like strands leading to thicker cables and finally to the spinal cord itself. The dendrites interweave so intricately that even with a microscope it is nearly impossible to discern where one ends and another begins. I liken the sight to standing on the edge of a forest on a winter day. Before me marches a line of several hundred trees, each thrusting black lengths of snow-laced branches up and out. If all those trees could be compressed together into a few square yards, with their branches intact and somehow filling up the spaces without actually touching each other, the resulting image would resemble a nerve bunch in the body.

A great debate raged in neurophysiology for decades: Do the branches, or dendrites, actually touch? In the electrical wiring of a home, of course, every live wire is joined by a wire nut to every other wire, so the system is, like plumbing, a completely closed loop. But in the body each of the twelve billion neurons stops just short of its neighbouring neurons, forming a precise gap called a synapse.

The synapse allows staggering complexity. Take just one motor neuron controlling one muscle fibre in one hand. Along the length of that one nerve cell, at thousands of separate points, knobs from other neurons form synapses. (On a large motor nerve ten-thousand separate contacts are made, and a brain neuron may have as many as eighty-thousand.) If a signal stimulates that motor nerve into action, immediately thousands of other nerve cells in the vicinity are put on alert. The single cells pictured in the biology texts extend to every square millimetre of skin, every muscle, every blood vessel, every bone – they have total saturation.

I want to move my hand. Is the stimulus from the brain strong enough to contract the muscle? How many muscle fibres are necessary for the appropriate strength? Are opposing muscles properly inhibited? The single nerve carries all these electrical messages, up to one-thousand separate impulses a second, with an appropriate pause between each. Every impulse is monitored and affected by all the ten-thousand synaptic connections along the path. The click-click-click auditory image of muscular action, then, is a kindergarten concept. Actually a stupendous crackling wildness surges in all of us at every moment.

Should we do something to alleviate the incessant hysteria of communication? Should I rest from my typing to allow my finger neurons to recover from their frenetic activity? To the contrary, our bodies seem to require an incredible volume of stimuli. Experimental subjects have deprived their bodies of normal daily stimuli with disastrous results. Some have locked themselves in dark, padded boxes; some have floated blindfolded and motionless in a tank of warm water. If nature abhors a vacuum, the brain abhors silence. It begins to break down, quickly filling the void with hallucinations. The volunteer begs for relief after a few hours – he cannot stay sane without the stimuli.

*

The brain cannot directly order each decision in the body – that would defy the management principle of delegation. Instead, a rigid, consistent reflex system handles many situations.

When I tap the tendon below a patient's knee, his leg flies up toward my face until muscle tension stops it. I tap it again, this time after telling the patient to stifle the reflex. He fails; the leg recoils anyway. What sinister force in his tendon dares to oppose his brain? It is simply a built-in protection system. Small, spindle-shaped structures, embedded near that tendon, stretch with the muscles, alerting nerve fibres to hurry the message to the spinal cord. Normally (in fact, almost always except in the case of the doctor's reflex test), sudden tension in that tendon means his leg has just been burdened by a heavy weight. Usually this happens when he is about to fall, and the process of stumbling triggers a reflex that automatically straightens his leg. The brain delegates such protections to the reflex arc. A reflex is built-in, dependable.

It shows good management principles, this delegation

to sneeze, cough, swallow, salivate, and blink. Blink. I have already mentioned the tragic blindness that afflicts leprosy patients who have lost eyelid reflexes. Nothing alerts them when the cornea has dried out and needs a lubricating blink. We can sometimes prevent the blindness by teaching the patients to blink. One would think patients with eyesight at stake would be eager learners, but conscious reflexes are not so simple. Patients must be trained with placards and stopwatches, drilled, scolded, praised, cajoled. The advanced brain informs them it cannot be bothered with something so elementary as a reflex (who would force a sophisticated IBM computer to count to ten every thirty seconds?) Some patients do not learn, and their eyes eventually dry out.

*

Some functions, though, do not fit the rigid, robotlike response of reflex. In the brain stem lies the next level of guidance, the subconscious regulators of breathing, digestion, and heart action. These need more attention than reflexes: simply breathing entails the cooperation of ninety chest muscles. And, the body's requirements change quickly; for example, heartbeat and breathing spurt wildly when I race up the stairs.

Highest of all in the hierarchy of the nervous system are the cerebral hemispheres of the brain, the holy of holies of the body − most protected by bone, most vulnerable to injury if the protection is ever breached. There ten billion nerve cells and one hundred billion glia cells (which provide the biological batteries for brain activity) float in a jellied mass, sifting through information, storing memories, creating consciousness. In the brain lies our proclivity to evil and rage as well as our impetus toward purity and love.

Already researchers can control rage, can, with a

transmitter implanted in the brain of a charging bull, electrically switch him into a harmless, cuddly pet. Some love to take complex notions, like romantic love, altruism, or an idea of God, and smilingly explain them away in terms of potassium ions and chemical balances and memory/association cells in the brain. But they get nowhere. How do I know that the idea of God isn't merely a series of electrical impulses in my brain? Answer: How do I know that electrical impulses are not God's chosen device for communicating to me a spiritual reality that could not otherwise be known?[2]

*

The hierarchy seems neatly ordered. But one messy problem keeps popping up, throwing a pencil into the well-lubricated gears of motion. The final decision, the localized "will" that controls muscles and movement, resides not in the magnificent crevasses of the brain, but in the humble, singular nerve cell or neuron that controls the muscle fibres. Sir Charles Sherrington discovered this discomfiting feature and grandiosely labelled it "the final common path."

The cell body of this neuron receives a spray of impulses from surrounding nerve centres. It stays alert to muscle tension, the presence of pain, the action of opposing muscles, the degree of strength required for any given activity, the frequency of stimuli, the oxygen available, the body temperature, the fatigue factor. Orders from the brain flood in: lift your arm – the box is heavy, so be ready to enlist a squadron of motor units. But after all the signals have accumulated in a giant contradictory pool of advice and recommendations, the motor neuron itself, down in the spinal cord, decides whether to contract or relax. It, after all, is best equipped for such a decision, being in intimate contact with thousands of local synapses as well as the brain.

Professor Bullock of the University of California in San Diego sums up the process: "The degrees of freedom available even at this low level can provide for an almost unlimited degree of complexity." Now that we have figured out the sequenced hierarchy of the body, it reduces to the simple fact that the neuron does whatever it thinks is best. Who said nature is not a democracy? Particle physicists have been telling us that for decades, and now our brain and its agents confirm the fact.

Only the "final common path" can decide between incompatible commands and reflexes, and we should be glad. I stand on a cliff on one of the sheer granite hulks in the Rocky Mountains. Ahead of me, just beyond my reach, is a delightful wildflower I have never seen before. I reach forward, peering through my camera under instructions from my brain, after carefully planting my feet. My closeup lens is within inches of the wildflower when suddenly a string is jerked and, like a marionette, I tip backward away from the flower. My heart is pounding, and I look around to see who interrupted my photography. No one is there, save a raucous, scolding jay.

Ever since I peered over the edge of the cliff to the ravine two thousand feet below, my cells have been chemically flooded with a heightened awareness of the potential danger. My conscious brain wanted a picture of the flower; my subconscious reflexes picked up a slight, precarious tilt in the balancing organs of my ear and shortcircuited the orders, by sending *urgent* messages directly to the nerve cells that controlled the muscles, abruptly yanking me backward.

The same saving rebellion takes over when I walk barefoot around the grounds of the Carville Hospital. (I am a great advocate of barefootedness, believing it makes for healthy, strong feet and opens up a whole world of sensation and awareness of the ground I walk on.) If I step on a thorn, my foot will stop in mid-step, pulling itself back even before the pain registers in my brain. But if I were escaping a burning plane, my cells would know that the brain was calling on them to bear some unusual stresses to prevent much more traumatic stress. I could then step on a flaming shard of metal because the normal reflex would be short-circuited for the sake of the more pressing goal of escape.*

The nervous system's hierarchy serves my sense of survival. Sometimes my brain overrules; sometimes it delegates. Always the results of its orders depend on the local, autonomous cell – the final common path.

---

* Jim Corbet, who wrote about India, described one remarkable case of will power overcoming pain in a time of stress. Upon examining the scene of a tiger attack, he found that a lady had gripped a tree branch so resolutely that the skin of her hands stayed on the branch while the tiger tore her body from it.

24
GUIDANCE

To will one thing, then, can only mean to will the Good,
because every other object is not a unity, and the will that
only wills that object, therefore, must become double-minded.
                                    SOREN KIERKEGAARD

We have charted the hierarchy within the body: cell to
neuron to reflex to conditioned reflex to brain stem to higher
brain, then back down the final common path to the
controlling neuron. Despite the complex interactions of
thousands of synapses, the system shows a basic simplicity
of design. It combines freedom with cooperation. Actions
as ordinary as squashing a mosquito or photographing a
flower tax the full capacity of this amazing system.

I cannot imagine a more striking parallel to the network
of communication that unites the members of Christ's
Body. All of us have declared allegiance to the Head, who
is Christ. But God, with His deep, implicit regard for
freedom, has left the final choice of action to individuals
who are as fully independent as the final common path.

The body offers one obvious lesson: all levels of
communication are important. Life would be hopelessly
complex if my brain had to give conscious orders for every
muscle contraction. As I walk to work in the monring, I
am free to think about my patients or about the birds

chirping in the branches. My legs need no conscious direction; their muscles follow the sequence of reflex activity that has been programmed into them. Motor units rest in turn rather than all at once, so my action can be continuous rather than sporadic. My neurons, alert to all other parts of my body, will slow my pace if my heart complains or take immediate action if I stumble.

A healthy body is a beautiful, singing harmony between the central nervous system and the tissues it controls. Yet in all this harmony every neuron must determine its own action based on the many impulses that come in. The microscopic computer in each nerve cell gauges my intentions, consults other muscles, analyses hormones, energy availability and the inhibition of fatigue or pain, and fires a yes or no order to its muscle group.

Think of yourself as a motor unit in Christ's Body, one of millions. How do you decide how and when to act? What is true guidance? Should a "higher" impulse always supersede a "lower" impulse?

*

Each cell's most numerous and immediate connections link it to local neurons. Some feed in from other motor cells, some from pain cells, pressure cells, temperature cells, muscle tone cells. All of these transmit waves of data which inform the individual neuron how to act in community. I believe that God has similarly delegated certain controls to the local church. How should the church respond to a decaying inner city? To the increasing pressures tearing families apart? To a disastrous flood? God has laid out principles governing the response of the whole Body, but He has also designated that local groups of His followers determine the role of each individual cell.

The Bible lists the various spiritual gifts that should be

used in organizing a hierarchy among the local members. Interestingly, when it outlines church offices, it does not recommend seeking out technically skilled people. There are no suggestions that a leader be a good manager or a sharp accountant or even show leadership potential. The essential qualities are spiritual qualities: How committed are they to God? Can they control their own temperaments? What are their families like? The key ingredient for all church offices listed in the Bible is not ability but loyalty. God seems to say, I will work with any people you give me as long as they are loyal. Having entrusted us with freedom, He needs leaders among us who are prone to exercise that freedom by aligning it with His will. A skilled but disloyal cell may initiate a wonderfully impressive action, but, like the spastic muscle, it will be useless unless it matches the body's needs.

Those of us raised in Western democracies, which highly value autonomy and freedom, readily respond to the image of the body as a composite democracy with the ultimate decision resting in individual cells. But that is only part of the story. As Bishop Lesslie Newbigin has stated it, "The goal of His [God's] purpose is not a collection of individual spirits abstracted one by one. . . . Such a thought is irreconcilable with the biblical view of God, of man, and of the world. The redemption with which He is concerned is both social and cosmic, and therefore the way of its working involves at every point the re-creation of true human relationships and of true relationship between man and the rest of the created order. Its centre is necessarily a deed wrought out at an actual point in history and at a particular place. Its manner of communication is through a human community wherein men are reborn into a new relation one with another, and become in turn the means of bringing others into that new relationship."[1] Often

God speaks to us not just through a direct approach to our own souls, but through fellow members of His Body. That very process binds us together with them.

Some Christian leaders have developed this mutual relationship to other cells in challenging ways. John Wesley's "Methodists" got their name from organized methods of making individuals responsible to others. In regular bands that met weekly, each member would answer to the group, "Have you faced temptations this week? Did you give in? How did you spiritually grow this week? What in your life needs prayer?" Early Methodists took seriously the chain of command which, in Christ's Body as in physical bodies, extends horizontally as well as vertically.

Daily each of us as individual cells face a myriad of choices: what to have for breakfast, what radio station to listen to, what toothpaste to use, which neighbours to see, which phone calls to make. Beyond these trivia are numerous ethical choices: How can I love my neighbour as myself? Is it wrong for me to use this excess income on a new shirt? How scrupulous should I be on my income tax return? How do I get God's guidance on these decisions? Psychiatrists can give many examples of well-meaning religious people who have been paralysed by just such perplexing questions. Our brains can be so occupied sorting through the blizzard of information that our response is helpless inactivity.

For this reason, I think, the Bible encourages us to ground ourselves in contact with God and His Word so thoroughly that our Christian actions become like reflexes to us. If I must decide whether to tell the truth in the face of every situation, my life is hopelessly complex. But if I have a reflex of truthfulness that responds without orders from higher up, I can learn to "walk" as a Christian without having to think about each individual step.

Paul summarized the process of our being imprinted with proper spiritual reflexes in this passage: "Do not conform any longer to the pattern of this world, but be transformed by the renewing of your mind. Then you will be able to test and approve what God's will is – His good, pleasing and perfect will." (Romans 12:2) From there he goes on to the first full mention in the New Testament of the body analogy, followed by a list of abrupt commands telling what God's will includes: Hate what is evil; cling to what is good. Honour one another above yourselves. Share with God's people who are in need. Don't be proud. Live at peace with everyone.

Paul never dwells on subtle psychological distinctions or explores all the family and sociological factors that would make such obedient behaviour difficult. He doesn't coax us toward right living. He simply states what it is and admonishes us to "renew our minds". I would paraphrase that as "reminding individual cells of their new identity in Christ". We have a tendency to forget, to substitute our educated selves for Christ as the Head. Paul recommends a process of mental purging, a conscious identification with the hierarchy God has set up.

*

Quite often I meet Christians who tend to wear their spirituality as an aura of otherworldliness. According to some, the most spiritual Christian is one who confidently asserts, "God told me it's time to buy a new dress," or "I'm positive God wants our church to use our money this way." "God told me" can become a casual manner of speech. Actually, I believe most of what God has to say to me is already written in the Bible and the onus is on me to diligently study His will revealed there. For most of us, mysterious direct messages from a hotline to God are not the ordinary ways of discerning His will. Guidance mediated through circumstances or modified on the advice of wise Christian friends, though it may seem less spectacular, is not at all inferior.

College graduates agonize over what decisions to make for the future, waiting for God to alert them with a jolting, custom-made plan dropped into their laps. In the Bible, God indeed employed the supernatural means of angels and visions and the like to convey His will. But, if you look closely at those incidents, you will note that few of them came in response to a prayer for guidance. They were usually unrequested and unexpected.

Consider the oft-cited Macedonian call of Paul as an example. Spectacular, yes, since a vision of a man beckoned Paul to change his travel plans and head toward Macedonia. Note carefully, however, that the vision prompted Paul to *change* his plans. You would expect Paul to plan his future in a godly way, but this incident indicates that Paul had gone off on his missionary journey without any visions or inner voices from the Spirit. Most likely, he looked over the situation and chose the route that seemed the most sensible. But the Holy Spirit wanted him to go into a whole new region – and so intervened spectacularly. It was exceptional guidance, obviously not the kind Paul normally relied on.

When grasping for analogies to describe the growing faith of the individual believer, Paul often turned to athletics: running, boxing, wrestling. Athletes demonstrate well the discipline that can train the body toward predictable, dependable actions. Pete Rose can count on his muscles springing into action to snare a screaming line drive because he has built that response into the neurons through hours of practice. An athlete's body knows what his mind wants and is equipped and experienced to effect that desire.

Similarly, the individual Christian would better spend his time working on practical, daily obedience to what God has already revealed rather than fervent searches for some magic secret, as elusive as the Holy Grail.

*

After stressing guidance from community and from trained reflexes, I must quickly add that each individual neuron does have a direct access to the brain. Although the pathway may not often be used in spectacular ways, it is present, and its synaptic connection can provide stirring, life-changing moments.

Such an experience happened to me my first year in India. I had had a general feeling I was supposed to be a missionary, so after my medical training I agreed to journey to India, the country of my birth. When the medical college first proposed my going, I stipulated a one-year contract because I was still tentative about my whole future. I went and taught, performed surgery, and filled whatever daily functions came up on the hospital staff.

Then, after being there a few months, I visited Dr Robert Cochrane, a renowned skin specialist, at the leprosy sanatorium in Chingleput a few miles south of Madras. My own hospital did not admit leprosy patients, and I had seen none professionally. Dr Cochrane showed me around the grounds of his hospital, nodding to the patients who were squatting, stumping along on bandaged feet, or following us with their unseeing, deformed faces. Gradually my nervousness (a result of my childhood memories) melted into a sort of professional curiosity, and my eyes were drawn to the hands of the patients.

Hands waved at me and stretched out in greeting. I study hands as some people study faces – often I remember them better than faces. But these were not the exquisite paradigms of engineering I had studied in medical school. They were twisted, gnarled, ulcerated stumps. Some were stiff like metal claws. Some were missing fingers. Some hands were missing altogether.

Finally I could restrain myself no longer. "Look here, Bob," I interrupted his long discourse on skin diseases, "I don't know much about skin. Tell me about these hands. How did they get this way? What do you do about them?"

Bob shrugged and said, "Sorry, Paul, I can't tell you. I don't know."

"Don't know!" I responded with obvious shock and amazement. "You've been a leprosy specialist all these years and you don't know? Surely something can be done for these hands!"

Bob Cochrane turned on me almost fiercely, "And whose fault is that, if I may ask – mine or yours? I'm a skin man – I can treat that part of leprosy. But you are a bone man, the orthopaedic surgeon!" More calmly, with sadness in his voice, he went on to tell me that not one orthopaedic surgeon had yet studied the deformities of the fifteen million leprosy victims in the world.

As we continued our walk, his words sank in to my mind. Far more people were afflicted by leprosy than those deformed by polio, or those mangled by motorcar accidents world-wide. But not one orthopaedist to serve them? Cochrane told me why he thought that was true: simple prejudice. Leprosy was surrounded by an aura of black magic. Most doctors would not get close to the leprosy patients. The few who did were idealistic or were priests and missionaries.

A few moments later I noticed a young patient sitting on the ground trying to take off his sandal. His disabled hands would not cooperate as he attempted to wedge the sandal strap between his thumb and the palm of his hand.

He complained that he could never grasp things – they always slipped from his hand. On sudden impulse, I moved toward him. "Please," I asked in Tamil, "may I look at your hands?"

The young man arose and, smiling, thrust his hands forward. I held them in mine, almost reluctantly. I traced the deformed fingers with my own and studied them intently. Finally, I pried his fingers open and placed my hand in his in a handshake grip. "Squeeze my hand," I directed, "as hard as you can."

To my amazement, instead of the weak twitch I had expected to feel a sharp intense pain raced through my palm. His grip was like a vice, with fingers digging into my flesh like steel talons. He showed no paralysis – in fact I cried out for him to let go. I looked up angrily but was disarmed by the gentle smile on his face. He did not know he was hurting me. And that was the clue. Somewhere in that severely deformed hand were powerfully good muscles. They were obviously not properly balanced, and he could not feel what force he was using. Could they be freed?

I felt a tingling as if the whole universe was revolving around me. I knew I had arrived in my place.

That single incident in 1947 changed my life. From that instant I knew my calling as surely as a cell in my body knows its function. Every detail of that scene – the people standing around the grounds, the shade of the tree, the questioning face of the patient whose hand I was holding – is still etched into my mind. It was my moment, and I had felt a call of the Spirit of God. I was made for that one moment in Chingleput, and I knew when I returned to my base I would have to point my life in a new direction. I have never doubted it since.

25
A PRESENCE

The Holy Spirit is the force in the straining muscles of an arm, the film of sweat between pressed cheeks, the mingled wetness on the backs of clasped hands. He is as close and unobtrusive as that, and as irresistibly strong.

JOHN V. TAYLOR

As a junior doctor on night duty in a London hospital I called on eighty-one-year old Mrs Twigg. This spry, courageous woman had been battling with cancer of the throat, but even with a raspy, hoarse voice she remained witty and cheerful. She had asked that we do all we could medically to prolong her life, and one of my professors removed her larynx and the malignant tissue around it.

Mrs Twigg seemed to be making a good recovery until about two o'clock one morning when I was urgently summoned to her ward. She was sitting on the bed, leaning forward, with blood spilling from her mouth. Wild terror filled her eyes. Immediately I guessed that an artery back in her throat had eroded. I knew no way to stop the bleeding but to thrust my finger into her mouth and press on the pulsing spot. Grasping her jaw with one hand, I explored with my index finger deep inside her slippery throat until I found the artery and pressed it shut.

Nurses cleaned up around her face while Mrs Twigg

recovered her breath and fought back a gagging sensation. Fear slowly drained from her as she began to trust me. After ten minutes had passed and she was breathing normally again, with her head tilted back, I tried to remove my finger to replace it with an instrument. But I could not see far enough back in her throat to guide the instrument, and each time I removed my finger the blood spurted afresh and Mrs Twigg panicked. Her jaw trembled, her eyes bulged, and she forcefully gripped my arm. Finally, I calmed her by saying I would simply wait, with my finger blocking the blood flow, until a surgeon and anaesthetist could be summoned from their homes.

We settled into position. My right arm crooked behind her head, supporting her. My left hand nearly disappeared inside her contorted mouth, allowing my index finger to apply pressure at the critical point. I knew from visits to the dentist how fatiguing and painful it must be for tiny Mrs Twigg to stretch her mouth open wide enough to surround my entire hand. But I could see in her intense blue eyes a resolution to maintain that position for days if necessary. With her face a few inches from mine, I could sense her mortal fear. Even her breath smelled of blood. Her eyes pleaded mutely, "Don't move – don't let go!" She knew, as I did, if we relaxed our awkward posture, she would bleed to death.

We sat like that for nearly two hours. Her imploring eyes never left mine. Twice during the first hour, when muscle cramps painfully seized my hand, I tried to move to see if the bleeding had stopped. It had not, and as Mrs Twigg felt the rush of warm liquid surge up in her throat she gripped my shoulder anxiously.

I will never know how I lasted that second hour. My muscles cried out in agony. My fingertip grew totally numb. I thought of rock climbers who have held their fallen partners for hours by a single rope. In this case the cramping four-inch length of my finger, so numb I could not even feel it, was the strand restraining life from falling away.

I, a junior doctor in my twenties, and this eighty-one-year old woman clung to each other superhumanly because we had no choice – her survival demanded it.

The surgeon came. Assistants prepared the operating room, and the anaesthetist readied his chemicals. Mrs Twigg and I, still entwined together in our strange embrace, were wheeled into the surgery room. There, with everyone poised with gleaming tools, I slowly eased my finger away from her throat. I felt no gush of blood. Was it because my finger could no longer feel? Or had the blood finally clotted firmly after two hours of pressure?

I removed my hand from her mouth and still Mrs Twigg breathed easily. Her hand continued to clutch my shoulder and her eyes stayed on my face. But gradually, almost imperceptibly at first, the corners of her bruised, stretched lips turned slightly up, forming a smile. The clot had held. She could not speak – she had no larynx – but she did not need words to express her gratitude. She knew how my muscles had suffered; I knew the depths of her fear. In those two hours in the slumbrous hospital wing, we had become almost one person.

*

As I recall that night with Mrs Twigg, it stands almost as a parable of the conflicting strains of human helplessness and divine power within us. In this case, my medical training counted very little. What mattered was my presence and my willingness to respond by reaching out and contacting another human being.

Along with most doctors I know, I often feel inadequate in the face of real suffering. Pain strikes like an earthquake, with crushing suddenness and devastation. A woman feels a small lump in her breast, and her sexual identity begins to crumble. A child is stillborn, and the mother wails in anguish: "Nine months I waited for this! Why do so many mothers abort their babies while I would give my life to have a healthy one?" A young boy is thrown through the windshield of a car, permanently scarring his face. His memory flickers on and off like a faulty switch – doctors, ever cautious, can't offer much hope.

When suffering strikes, those of us standing close by are flattened by the shock. We fight back the lumps in our throats, march resolutely to the hospital for visits, mumble a few cheerful words, perhaps look up articles on what to say to the grieving.

But when I ask patients and their families, "Who helped you in your suffering?" I hear a strange, imprecise answer. The person described rarely has smooth answers and a winsome, effervescent personality. It is someone quiet, understanding, who listens more than talks, who does not judge or even offer much advice. "A sense of presence." "Someone there when I needed him." A hand to hold, an understanding, bewildered hug. A shared lump in the throat.

We want psychological formulas as precise as those techniques I study in my surgery manuals. But the human psyche is too complex for a manual. The best we can offer is to be there, to see and to touch.

Several themes have recurred throughout this book: the need to serve the Head loyally, the unobtrusive nature of the Body's firm skeleton, the softness and compliancy of the skin, and the healing activity of Christ's Body. Taken together, these provide a sense of presence to the world – God's presence.

Sometimes I, as a member of Christ's Body, feel as if

I am back in the room with Mrs Twigg. All my parts – bone, muscle, blood, brain – collaborate beautifully to allow me to stave off certain death in my patient. Yet I must also fend off a sense of helpless futility. The most I can do is dam the flow of blood for a short while, delaying the further invasion of her terminal cancer. I wish, instead, for a miracle.

*

Is God's plan to possess the earth through a Body composed of frail humans adequate in light of the sheer enormity of the world's problems? Such a question deserves the full treatment of a book much longer and wiser than this one. I can, however, capture a glimpse of God's style of relating to our planet by reviewing the progressive metaphors He has given us.

All language about God is, of course, symbolic. "Can one hold the ocean in a teacup?" Joy Davidman asked. Words, even thoughts, cannot carry Godness. In the Old Testament, symbols for God most often expressed His "otherness". He appeared as a Spirit, so full of light and glory that one who approached Him was struck dead or returned with an unhuman glow. Moses saw only God's back; Job heard Him from a whirlwind; the Israelites followed His shekinah glory cloud.

Is it any wonder that the Jews, accustomed to such mystery and afraid to say aloud or write the name of God, recoiled at the claims of Jesus Christ? "Anyone who has seen me has seen the Father," Jesus said (John 14:9), words that grated harshly on Jewish ears. He had, after all, spent nine foetal months inside a young girl and had grown up in a humble neighbourhood. In Chesterton's words, "God who had been only a circumference was seen as a centre; and a centre is infinitely small."[1] Visibly at least, he seemed too much like any other human. Their suspicions were confirmed when He succumbed to death. How could God be contained inside the flesh and blood of humanity? How could God die? Many still wonder, long after a resurrection that convinced and ignited His followers.

But Jesus departed, leaving no body on earth to exhibit the Spirit of God to an unbelieving world – except the faltering, bumbling community of followers who had largely forsaken Him at His death. *We* are what Jesus left on earth. He did not leave a book or a doctrinal statement or a system of thought; He left a visible community to embody Him and represent Him to the world. The seminal metaphor, Body of Christ, hinted at by Christ and fully expanded by Paul could only arise *after* Jesus Christ had left the earth.

The apostle Paul's great, decisive words about the Body of Christ were addressed to congregations in Corinth and Asia Minor that, in the next breath, he assailed for human frailty. Note that Paul, a master of simile and metaphor, did not say the people of God are "like the Body of Christ." In every place he said we *are* the Body of Christ. The Spirit has come and dwelt among us, and the world knows an invisible God mainly by our representation, our "enfleshment," of Him.

"The Church is nothing but a section of humanity in which Christ has really taken form," said Bonhoeffer.[2] Too often we shrink from both clauses of that summary. Dismayed, we blast ourselves for continuing to manifest our flawed humanity. Disheartened, we in practice, if not in faith, deny that Christ really has taken form in us.

Three dominant symbols – God as a glory cloud, as a Man subject to death, and as a Spirit melding together His new Body – show a progression of intimacy, from fear to shared humanity to shared essence. God is present in us, uniting us genetically to Himself and to each other.

Where is God in the world? What is He like? We can no longer point to the Holy of Holies or to a carpenter in Nazareth. *We* form God's presence in the world through the indwelling of His Spirit. It is a heavy burden.

*

After World War II German students volunteered to help rebuild a cathedral in England, one of many casualties of the Luftwaffe bombings. As the work progressed, debate broke out on how to best restore a large statue of Jesus with His arms outstretched and bearing the familiar inscription, "Come unto Me." Careful patching could repair all damage to the statue except for Christ's hands, which had been destroyed by bomb fragments. Should they attempt the delicate task of reshaping those hands?

Finally the workers reached a decision that still stands today. The statue of Jesus has no hands, and the inscription now reads, "Christ has no hands but ours."

I show you a mystery: "In him you too are being built together to become a dwelling in which God lives by his Spirit" (Ephesians 2:22).

NOTES

PREFACE

[1]G. K. Chesterton, *St Francis of Assisi* (Garden City, N.Y.: Doubleday & Co., 1957) p. 31.

CHAPTER 2

[1]Annie Dillard, *Pilgrim at Tinker Creek* (New York: Harper's Magazine Press, 1974) p. 94.
[2]Lewis Thomas, *The Medusa and the Snail* (New York: Viking Press, 1979) pp. 155–57.

CHAPTER 3

[1]Frederick Buechner, *Telling the Truth* (New York: Harper & Row, 1977) pp. 57–8.
[2]C. S. Lewis, *God in the Dock* (Grand Rapids, Mich.: Eerdmans, 1979) p. 62, and London: Fount Paperbacks.

CHAPTER 7

[1]Statistics compiled by World Vision International.

CHAPTER 10

[1]G. K. Chesterton, *Orthodoxy* (Garden City, N.Y.: Doubleday & Co., 1959) p. 95.
[2]Ibid., p. 58.

CHAPTER 12

[1]Adapted from Mont Smith, "The Temporary Gospel", *The Other Side* (November-December, 1975).

CHAPTER 13

[1]Merton P. Strommen, *Five Cries of Youth* (New York: Harper & Row, 1974) p. 76.

CHAPTER 16

[1] R. J. Christman, *Sensory Experience* (Scranton, Pa.: Intext Educational Publishers, 1971) p. 359.

CHAPTER 17

[1] Ashley Montagu, *Touching* (New York: Columbia University Press, 1971) p.30.
[2] Ibid., p. 82.

CHAPTER 21

[1] Jonathan Miller, *The Body in Question* (New York: Random House, 1978) p. 310.
[2] Charles Williams, *The Descent of the Dove* (London: Longmans, Green and Company, 1939) p. 31.
[3] Ibid., p. 57.

CHAPTER 23

[1] Robert Galambos, *Nerves and Muscles* (Garden City, N.Y.: Doubleday & Co., 1962) p. 23.
[2] Robert Farrar Capon, *The Third Peacock* (Garden City, N.Y.: Doubleday & Co., 1971) p. 48.

CHAPTER 24

[1] Lesslie Newbigin, *The Household of God* (New York: Friendship Press, 1954) pp. 109–110.

CHAPTER 25

[1] G. K. Chesterton, *The Everlasting Man* (Garden City, N.Y.: Image Books, 1955) p. 174.
[2] Dietrich Bonhoeffer, *Ethics* (London: SCM Press, 1971) p. 64.

BIBLIOGRAPHY

Andrews, Michael. *The Life That Lives on Man*. New York: Taplinger Publishing Co., 1976.

Bevan, Edwyn. *Symbolism and Belief*. London: Fontana, 1938.

Bourne, Geoffrey H., ed. *The Biochemistry and Physiology of Bone*. New York: Academic Press, 1959.

Bullock, Theodore Holmes. *Introduction to Nervous Systems*. San Francisco: W. H. Freeman and Co., 1977.

Carlson, Anton J.; Johnson, Victor; Cavert, H. Mead. *The Machinery of the Body*. Chicago: The University of Chicago Press, 1976.

Carterette, Edward C., and Friedman, Morton P., ed. *Handbook of Perception*, Volume 3. New York: Academic Press, 1973.

Christman, R. J. *Sensory Experience*. Scranton, Pa.: Intext Educational Publishers, 1971.

Cole, Alan. *The Body of Christ*. Philadelphia: Westminster Press, 1964.

Communications Research Machines. *Biology Today*. Del Mar, California, 1972.

Curtis, Helena. *Biology: The Science of Life*. New York: Worth Publishers, 1968.

Drummond, Henry. *Natural Law in the Spiritual World*. London: Hodder and Stoughton, 1887.

Eckstein, Gustav. *The Body Has a Head*. New York: Harper & Row, 1969.

Espenschade, Anna S. and Eckert, Helen M. *Motor Development*. Columbus, Ohio: Merrill Books, Inc. 1976.

Galambos, Robert. *Nerves and Muscles*. Garden City, N.Y.: Doubleday & Co., 1962.

Herbert, Don and Bardossi, Fulvio. *Secret in the White Cell*. New York, N.Y.: Harper & Row, 1969.

Huxley, Thomas Henry. *The Crayfish*. Cambridge, Mass.: MIT Press, 1880.

Kenshalo, Dan R. *The Skin Senses*. Springfield, Ill.: Charles C. Thomas, 1969.

Langley, LeRoy. *Physiology of Man*. New York: Van Nostrand Reinhold Co., 1971.

Lenihan, John. *Human Engineering*. New York: George Braziller, 1974.

Miller, Jonathan. *The Body in Question*. New York: Random House, 1978.

Minear, Paul S. *Images of the Church in the New Testament*. Philadelphia: Westminster Press, 1960.

Montagna, William. *The Structure and Function of Skin*. New York, N.Y.: Academic Press, 1956.

Montagu, Ashley. *Touching*. New York: Columbia University Press, 1971.

Nilsson, Lennart with Lindberg, Jan. *Behold Man*. Boston: Little, Brown and Co., 1973.

Nourse, Alan E. *The Body*. Alexandria, Virginia: Time-Life Books, 1964.

Robinson, John A. T. *The Body*. London: SCM Press, 1952.

Spearman, R. I. C. *The Integument*. London: Cambridge University Press, 1973.

Stevens, Leonard A. *Neurons: Building Blocks of the Brain*. New York: Thomas Y. Crowell Co., 1974.

Sutton, Richard L., Jr. *The Skin: A Handbook*. Garden City, N.Y.: Doubleday & Co., 1962.

Taylor, John V. *The Go-Between God*. London: SCM Press, 1974.

Tricker, R. A. R. and Tricker, B. J. K. *The Science of Movement*. New York: American Elsevier Publishing Co., 1967.

Von Frisch, Karl. *Man and the Living World*. New York: Harcourt, Brace & World, 1949.